THE PRACTICE OF ECONOMETRICS

INTERNATIONAL STUDIES IN ECONOMICS AND ECONOMETRICS

Volume 15

1. Harder T: Introduction to Mathematical Models in Market and Opinion Research With Practical Applications, Computing Procedures, and Estimates of Computing Requirements. Translated from the German by P.H. Friedlander and E.H. Friedlander. 1969.
2. Heesterman ARG: Forecasting Models for National Economic Planning. 1972.
3. Heesterman ARG: Allocation Models and their Use in Economic Planning. 1971.
4. Durdağ M: Some Problems of Development Financing. A Case Study of the Turkish First Five-Year Plan, 1963–1967. 1973.
5. Blin JM: Patterns and Configurations in Economic Science. A Study of Social Decision Processes. 1973.
6. Merkies AHQM: Selection of Models by Forecasting Intervals. Translated from the Dutch by M. van Holten-De Wolff. 1973.
7. Bos HC, Sanders M and Secchi C: Private Foreign Investment in Developing Countries. A Quantitative Study on the Evaluation of its Macro-Economic Impact. 1974.
8. Frisch R: Economic Planning Studies Selected and Introduced by Frank Long. Preface by Jan Tinbergen. 1976.
9. Gupta KL: Foreign Capital, Savings and Growth. An International Crosssection Study. 1983.
10. Bochove CA van: Imports and Economic Growth. 1982.
11. Bjerkholt O, Offerdal E (eds.): Macroeconomic Prospects for a Small Oil Exporting Country. 1985.
12. Weiserbs D (ed.): Industrial Investment in Europe: Economic Theory and Measurement. 1985.
13. Graf von der Schulenburg J-M, Skogh G (eds.): Law and Economics & The Economics of Legal Regulation. 1986.
14. Svetozar Pejovich (ed.): Socialism: Institutional, Philosophical and Economic Issues. 1987.
15. Heijmans RDH, Neudecker H (eds.): The Practice of Econometrics. 1987.

The Practice of Econometrics

Studies on Demand, Forecasting, Money and Income

edited by
Risto Heijmans and
Heinz Neudecker
(University of Amsterdam)

HB
139
.P7
1987
WEST

1987 **KLUWER ACADEMIC PUBLISHERS**
DORDRECHT / BOSTON / LANCASTER

Distributors

for the United States and Canada: Kluwer Academic Publishers, P.O. Box 358, Accord Station, Hingham, MA 02018-0358, USA
for the UK and Ireland: Kluwer Academic Publishers, MTP Press Limited, Falcon House, Queen Square, Lancaster LA1 1RN, UK
for all other countries: Kluwer Academic Publishers Group, Distribution Center, P.O. Box 322, 3300 AH Dordrecht, The Netherlands

Library of Congress Cataloging in Publication Data

```
The Practice of econometrics.

   (International studies in economics and econometrics ;
v. 15)
   Published in honor of Jan Salomon Cramer.
   Bibliography: p.
   Includes index.
   1. Econometrics.  2. Econometric models.
I. Neudecker, Heintz.  II. Heijmans, Risto.
III. Cramer, J. S. (Jan Salomon), 1928-
IV. Series.
HB139.P7  1987          330'.028        87-4054
ISBN 90-247-3502-5
```

ISBN 90-247-3502-5 (this volume)
ISBN 90-247-2730-8 (series)

Copyright

© 1987 by Martinus Nijhoff Publishers, Dordrecht.

All rights reserved. No part of this publication may be reproduced, stored in a retrieval system, or transmitted in any form or by any means, mechanical, photocopying, recording, or otherwise, without the prior written permission of the publishers,
Martinus Nijhoff Publishers, P.O. Box 163, 3300 AD Dordrecht, The Netherlands.

PRINTED IN THE NETHERLANDS

Contents

Preface VII

About the authors XI

A. Demand

1. R.W. Blundell and C. Meghir
   Engel curve estimation with individual data 3
2. S. Georgantelis, G.D.A. Phillips and W. Zhang
   Estimating and testing an almost ideal demand system 15
3. T. Lancaster, G. Imbens and P. Dolton
   Job separations and job matching 31
4. J. Muellbauer
   Estimating the intertemporal elasticity of substitution for consumption from household budget data 45
5. H. Theil
   Associated with an income distribution and a demand system is a multidimensional expenditure distribution 59

B. Money

6. E.L. Feige
   The theory and measurement of cash payments: a case study of the Netherlands 67
7. C.F. Manski and E. Goldin
   The denomination-specific demand for currency in a high-inflation setting: the Israeli experience 99

8. M. van Nieuwkerk
   In Search of 100 billion dollars 121
9. A.F. de Vos
   Forecasting the daily balance of the Dutch Giro 131

C. Income

10. A.H.Q.M. Merkies
    Inductive analysis from empirical income distributions ... 151
11. M.R. Ransom
    Economic growth and the size distribution of income; a
    longitudinal analysis 165

D. Methodology

12. A.P. Barten
    The coefficient of determination for regression without a
    constant term 181
13. R.D.H. Heijmans and H. Neudecker
    The coefficient of determination revisited 191
14. M.M.G. Fase
    Modelling multivariate stochastic time series for prediction:
    another look at the Lydia Pinkham Data 205
15. J.F. Kiviet and G. Ridder
    On the rationale for and scope of regression models in
    econometrics 223
16. D.S.G. Pollock
    The classical econometric model 247

Bibliography of Jan Salomon Cramer 263

Name Index 267

Subject Index 271

Preface

In the autumn of 1961 Jan Salomon ('Mars') Cramer was appointed to the newly established chair of econometrics at the University of Amsterdam. This volume is published to commemorate this event. It is well-known how much econometrics has developed over the period under consideration, the 25 years that elapsed between 1961 and 1986. This is specifically true for the areas in which Cramer has been actively interested. We mention the theory and measurement of consumer behaviour; money and income; regression, correlation and forecasting. In the present volume this development will be highlighted.

Sixteen contributions have been sollicited from scholars all over the world who have belonged to the circle of academic friends of Cramer for a shorter or longer part of the period of 25 years. The contributions fall broadly speaking into the four areas mentioned above. Theory and measurement of consumer behaviour is represented by four papers, whereas a fifth paper deals with a related area.

Richard Blundell and Costas Meghir devote a paper to the estimation of Engel curves. They apply a discrete choice model to British (individual) data from the Family Expenditure Survey 1981. Their aim is to assess the impact of individual characteristics such as income, demographic structure, location, wages and prices on commodity expenditure.

S. Georgantelis, Garry Phillips and W. Zhang examine the socalled Almost Ideal Demand System as introduced by Deaton and Muellbauer, and a simplified version which is also due to these authors. They then try to answer the question whether the use of the latter is justified. Their conclusion is that it is preferable to use the full AIDS which emerges here as the most appropriate vehicle for testing theoretical restrictions and obtaining elasticity estimates. The analysis is on a macrolevel. It uses British data for the period 1950–83.

Tony Lancaster, Guido Imbens and Peter Dolton contribute a paper in the field of job separation and job matching. Though this does not belong to the area of consumer behaviour proper, it is of particular relevance, because it deals with problems of job tenure and wages. Also this paper is empirically slanted. It analyzes British data on a micro level coming from a survey of university graduates of the year 1970 conducted in 1977.

John Muellbauer reflects on Cramer's contributions to the study of household behaviour. His paper is about the problem of estimating the intertemporal elasticity of substitution from household budget data. It emerges that the curvature of Engel curves also governs that elasticity. British data pertaining to 11 commodity groups and the 5 years from 1968 to 1972 are used to establish the result.

Henri Theil who was Cramer's superior at the Central Planning Office at the Hague in the late fifties addresses the theoretical problem of deriving a multidimensional expenditure distribution from the income distribution of the consumers. Three examples, all based on a lognormal income distribution are added.

The second main area is that of money and income. It is covered by six papers.

Edgar Feige contributes on the theory and measurement of cash payments, a case study of the Netherlands. Its theoretical point of departure is Fisher's exchange equation. The author follows the approach by Laurent to determine the payment velocity of currency.

Charles Manski and Ephraim Goldin examine the pattern of demand for Israeli currency between denominations and over time. The period which they consider is one of very high inflation, viz. 1969–81. Their model is also capable of forecasting currency demand specifically under high-inflation conditions. Another monetary paper is by Marius van Nieuwkerk. It is about the statistical discrepancy in the world balance of payments

The paper entitled 'In search of 100 billion dollars' unveils part of the mystery. The approach is non-technical, of a statistical-definitional nature.

To forecast the daily balance of the Dutch Postal Giro System is the task Aart de Vos set himself. His paper is of a statistical-methodological nature. It considers models for trends and seasonality, but also for calendar effects. Use is made of the Kalman filter technique.

There are two papers on income distribution. One by Arnold Merkies deals with the problem of how to find the (exact) shape of the size distribution of personal incomes. The author dismisses the Pareto distribution as a serious candidate, and discusses several alternatives. US data are used to illustrate the theoretical exposition.

Michael Ransom addresses the problem of possible causal relationship between economic development and the size distribution of personal incomes. He examines data from seven countries and subjects these to a longitudinal analysis. His dependent variable is income inequality as measured by the Gini coefficient or by the income share of the lowest quintile.

The fourth area covered in this volume is methodology. There are five papers devoted to the area.

Anton Barten contributes a paper on the definition and asymptotic behaviour of the coefficient of determination for regression without a constant term. He presents a general expression for the coefficient which covers that case and the one of regression with a constant term.

Risto Heijmans and Heinz Neudecker also study the coefficient of determination in linear regression models. They choose a different definitional starting point, which enables them to comprise linear regression models, with or without a constant, to be estimated by OLS or GLS. They give results on the asymptotic behaviour of the coefficient.

A paper by Martin Fase is looking into the problem of modelling multivariate stochastic time series with a view to forecasting. The study is mixed theoretical-practical, the theory being applied to advertising and sales data from US pharmaceutical industry. Both monthly and yearly data are analyzed, the first from the period 1954(1)–1960(6), the latter from the period 1907–1960.

Jan Kiviet and Geert Ridder discuss a methodological issue, viz. the rationale and scope of regression models in econometrics. Their paper pays attention to longstanding problems like the applicability of the models to cross-section and time-series situations.

Finally Stephen Pollock gives a thoroughgoing history of the classical econometric model.

We wish to thank the authors for their contributions, and the publishers for producing the volume. Thanks are also due to Marianne Coquillon, Andries Jansen and Hermance Mettrop of the Faculty of Actuarial Science and Econometrics for ther cooperation.

We sincerely hope that all readers of the volume will find some joy in it.

The Editors

Jan Salomon Cramer

About the authors

Anton P. Barten (The Coefficient of Determination for Regression without Constant term) is Professor of Econometrics at the Katholieke Universiteit Leuven, Belgium and Center for Operations Research and Econometrics, Louvain-la-Neuve, Belgium.
Richard Blundell (Engel Curve Estimation with Individual Data) is Professor of Economics at University College London, England.
Aart F. de Vos (Forecasting the Daily Balance of the Dutch Giro) is Senior Lecturer in the Department of Econometrics at the Free University Amsterdam, the Netherlands.
Peter J. Dolton (Job Separations and Job Matching) is Lecturer in Economics at the University of Hull, England.
Martin M.G. Fase (Modelling Multivariate Stochastic Time Series for Prediction) is deputy director of De Nederlandsche Bank N.V., Amsterdam, the Netherlands, chief of the Bank's Econometric Research and Special Studies Department and a part-time Professor of Marketing Analysis and Business Statistics at the Erasmus University, Rotterdam, the Netherlands.
Edgar L. Feige (The Theory and Measurement of Cash Payments: A Case Study of the Netherlands) is Professor of Economics at the University of Wisconsin, Madison, Wisconsin, U.S.A.
S. Georgantelis (Estimating and Testing an Almost Ideal Demand System) is a former Research Fellow in Economics at the University of Leeds, England.
Ephreim Goldin (The Denomination-Specific Demand for Currency in a High-Inflation Setting: the Israeli Experience) is currently employed as an Economist at the Currency Department of the Bank of Israel, Jerusalem, Israel.
Risto D.H. Heijmans (The Coefficient of Determination Revisited) is Senior Lecturer in the Department of Econometrics at the University of Amsterdam, the Netherlands.
Guido Imbens (Job Separations and Job Matching) is a Research Officer, Department of Economics and Commerce, University of Hull, England.

Jan F. Kiviet (On the Rationale for and Scope of Regression Models in Econometrics) is Lecturer in the Department of Econometrics at the University of Amsterdam, The Netherlands.

Tony Lancaster (Job Separations and Job Matching) is Professor of Econometrics, Department of Economics and Commerce, University of Hull, England.

Charles F. Manski (The Denomination-Specific Demand for Currency in a High-Inflation Setting: the Israeli Experience) is Professor of Economics at the University of Wisconsin, Madison, Wisconsin, U.S.A.

Costas H.D. Meghir (Engel Curve Estimation with individual Data) is Lecturer in Economics at University College London, England.

Arnold H.Q.M. Merkies (Inductive Analysis from Empirical Income Distributions) is Professor of Econometrics at the Free University, Amsterdam, the Netherlands.

John Muellbauer (Estimating the Intertemporal Elasticity of Substitution for Consumption from Household Budget Data) is Official Fellow at Nuffield College, Oxford, England.

Heinz Neudecker (The Coefficient of Determination Revisited) is Professor of Econometrics at the University of Amsterdam, the Netherlands.

Garry D.A. Phillips (Estimating and Testing an Almost Ideal Demand System) is Professor of Econometrics at the University of Leeds, England.

Stephen Pollock (The Classical Econometric Model) is Lecturer in the Department of Economics, Queen Mary College, London, England.

Michael R. Ransom (Economic Growth and the Size Distribution of Income: A Longitudinal Analysis) is Assistant Professor of Economics, Department of Economics at the University of Arizona, Tucson, Arizona, U.S.A.

Geert Ridder (On the Rationale for and Scope of Regression Models in Econometrics) is Lecturer in the Department of Econometrics at the University of Amsterdam, The Netherlands.

Henri Theil (Associated with an Income Distribution and a Demand System is a Multidimensional Expenditure Distribution) occupies the McKethan-Matherly Eminent Scholar Chair at the University of Florida's Graduate School of Business, Gainesville, Florida, U.S.A.

Marius van Nieuwkerk (In Search of 100 Billion Dollars) is Chief of the Balance of Payments Department of De Nederlandsche Bank N.V., Amsterdam, the Netherlands.

W. Zhang (Estimating and Testing an Almost Ideal Demand System) is Lecturer in Economics, Department of Management at the University of Fudan, Shanghai, Chinese People's Republic.

A. Demand

Engel curve estimation with individual data

RICHARD BLUNDELL and COSTAS MEGHIR

1. Introduction

In order to accurately assess the impact of important individual characteristics such as income, demographic structure, location, wages and prices on commodity expenditure it is most advantageous to work with survey data at the individual level. However, all surveys on individual or household expenditures cover purchases over a limited period of time. Indeed, the more accurate the measurement in terms of precise diary records the shorter the period over which the survey can be afforded and the more likely the occurrence of reported zero expenditures. Even where the expenditure in question relates to an item or group of items for which actual consumption is positive, infrequency of purchase or the 'durability' of the good may result in a recorded zero expenditure. Here we wish to develop a bivariate model for the joint determination of the purchase and expenditure decision. We shall focus on the particular example of a commodity which is continuously consumed but only infrequently purchased.

Although the use of individual survey data for Engel curve estimation formed a major part of earlier work on consumer behaviour (see Cramer (1964, 1971)) it is only recently that the issues raised relating to estimation of consumer demand functions using such data have attracted the popular attention they deserve. A major reason for this must be the dominance of the simple Tobit model (or censored univariate normal model) for data exhibiting zero values with finite probability for the dependent variable. Although clearly ideal for certain circumstances, the presence of zeros of the type described above in expenditure surveys would seem to require a more careful analysis. A number of papers have produced breakthroughs in this field notably the 'double-hurdle' model of Cragg (1971) subsequently applied to alcohol and tobacco expenditures by Atkinson, Gomulka and Stern (1983) and the infrequency-of-purchase model of Kay, Keen and Morris (1983) and Keen (1985). The central novelty of these developments over the Tobit model is the introduction of a bivariate process to describe jointly the discrete buy/non-buy decision and the conditional expenditure decision. This distinction is the central motivation for the models described in this paper. By concentrating on the case where purchases are infrequent while consumption is always positive, we conveniently sidestep the Tobit model where zero expenditures are exclusively determined by corner solutions. Where demand systems or Engel curves are estimated for 'grouped expenditures' (clothing, services, communications etc.) even at the individual level the majority of reported zero expenditures are likely to result from infrequency of purchase, not corner solutions, provided that expenditure is measured accurately. Corner solutions could, nevertheless, be introduced into the model described in this paper if required.

2. A model for infrequency of purchase and Engel curve estimation

As soon as it is recognised that reported expenditures in survey data may be contaminated by infrequency of purchase and therefore may not measure consumption accurately it is also the case that any measure of total consumption within a household will suffer measurement error. This is critical in the estimation of the parameters of utility consistent demand systems since it is usually assumed that total consumption is measured without error and demand analysis can proceed to examine the way in which consumption on specific commodities is allocated conditional on total consumption. It is for this reason that we focus in this paper on Engel curve estimation, that is the relationship between consumption, income and other characteristics, rather than worry about identifying all parameters of a complete demand system. Since we are unlikely to recover accurate price responses in cross-section survey data or short panels this does not appear to be a significant drawback. It does, however, allow us to concentrate on the probability-of-purchase model and its implications for econometric estimation and specification testing. Of course, under certain types of two-stage budgetting rules for household preferences, income-expenditure models of the kind estimated here can be used to directly recover parameters of the underlying demand system and consumer welfare function.

Closest in spirit to the approach developed here are the models of Kay, Keen and Morris (1984) and Deaton and Irish (1984). The infrequency of purchase models discussed in these papers have a purchase probability that although less than unity does not vary across individuals and is therefore determined independently of both the level of household income and other characteristics as well as the level of consumption itself. As is shown by Deaton and Irish (1984) this model is then observationally equivalent to a model with a constant probability of under-reporting and/or nonconsumption due to abstention (e.g. non-smokers). Our approach is to allow purchase probabilities to vary with income and other household characteristics so as to minimize the extent to which 'unobservable' characteristics are left to explain the distribution of purchases. This is not to abstract from the importance of the misreporting model but rather to separate it from the frequency-of-purchase model. For the types of commodities we consider in our empirical study (unlike the alcohol and tobacco example of Deaton and Irish (1984)) systematic under-reporting is unlikely to be an issue. Our general feeling is that observed consumption behaviour can be broken down into three broad categories. Those that are infrequently purchased but continuously consumed such as clothing, footwear, services, transport, etc. Those that are segmented between buyers and non-buyers such as tobacco and to a lesser extent alcohol and those that display the standard corner solution such as dissaggregated food expenditures, hours of work, etc. Although for many commodities or individuals these could overlap it may be useful to separate them for items of consumption that very clearly fall into one or another of the categories.

For commodities where consumption expenditure is observed, non-storable food items would be a good example, the Engel curve for some individual i may be written

$$c_i = g_i(x_i, \theta, u_i) \qquad (2.1)$$

where c_i is consumption and g_i is some function of observed characteristics x_i, preference parameters θ, and unobserved or unmeasured characteristics u_i. A convenient form for (2.1) when the commodity is continuously consumed across all individuals would be the log-linear specification

$$\log c_i = \theta' x_i + u_i \qquad (2.2)$$

where x_i would often contain log income terms as well as demographic characteristics and where u_i would be assumed independent of these x_i.

The infrequency-of-purchase model throws up the two serious problems for such a specification. Firstly, c_i is not observed. Rather the actual expenditure e_i in the survey period is observed, which may be zero and therefore may underestimate c_i or may be positive and therefore probably overestimate c_i since stocking-up is taking place. This, of course, is the central problem to be discussed in this paper. A second problem arises in log specification (2.2) since if consumption c_i were to be replaced by measured expenditure e_i the log transformation could not be taken when e_i was identically zero. As Cramer notes in his Empirical Econometrics text this is one of the prime reasons why transformations of the type (2.2) are not used in applied work on micro data. However, the infrequency-of-purchase model requires that c_i be treated as an unobservable latent variable and involves an additional probability of purchase term which allows an Engel curve like (2.2) to be adopted in estimation. Where the unobservable consumption is always positive, (2.2) provides a perfectly adequate specification and although not consistent on a system-wide basis with utility maximising theory may be ideally suited for modelling the consumption of certain individual commodities. Indeed, it has the advantage (over a shares model for example) of allowing a distribution for u_i which has no finite truncation points – the normal for example!

To be more formal we begin by stating the complete relationship between consumption c_i and expenditure e_i. Observed expenditure, e_i and consumption c_i are linked by the identity

$$E(e_i) \equiv E(e_i|Q_i>0) \Pr(Q_i>0) + E(e_i|Q_i\leq 0) \Pr(Q_i\leq 0) \qquad (2.3)$$

where $Q_i>0$ indicates that a purchase has taken place during the survey period. Noting that $E(e_i|q_i\leq 0) = 0$ and assuming that $E(e_i) = E(c_i)$ (a testable assumption, see Blundell and Meghir (1986)) we can now write

$$E(e_i|Q_i>0) \Pr(Q_i>0) = E(c_i) \qquad (2.3')$$

Consequently the relationship between e_i and c_i for individuals that are observed to purchase can be written as

$$e_i = c_i \exp(v_i)/p_i \qquad (2.4)$$

where $p_i = \Pr(Q_i>0)$ and where v_i is assumed to have some well behaved distribution such that $E\exp(v_i) = 1$. For households recording positive expenditures we have

$$\ln p_i + \ln e_i = \theta \cdot x_i + w_i \qquad (2.5)$$

where $w_i = u_i + v_i$.

For households with zero expenditures (2.5) is unobservable and we simply know that their behaviour occurred with probability $1 - p_i$.

Precisely how the determinants of the latent indicator Q_i and consequently p_i are chosen is to a large extent an empirical issue but they would probably depend on the relative time and money cost of making a purchase. In general we specify

$$Q_i = \alpha'z_i + \mu_i \qquad (2.6)$$

where $e_i>0$ if $Q_i>0$
and $e_i = 0$ if $Q_i\leq 0$.

where z_i is a vector of individual characteristics and $\mu_i \sim N(0,1)$. Whether μ_i can be assumed independent of w_i is yet another empirical question but clearly any model assuming independence should test such an assumption. Given some bivariate distribution for w_i and μ_i (bivariate normal may be a useful choice at the outset) even under independence deriving the sample likelihood and diagnostic statistics requires some care. In general the main problems which are considered in detail in Blundell and Meghir (1986b) relate to the Jacobian of transformation between w_i to e_i in (2.5) and the dependence of p_i on c_i. However, in the log model under independence the Jacobian (at least) is particularly simple as can be deduced from (2.5). The sample loglikelihood for a random drawing of H individuals is simply

$$\log L = \sum_+ [\log p(\alpha'z_i) + \log(f(\ln p_i(\alpha'z_i) + \ln e_i - \theta'x_i, \sigma))]$$
$$+ \sum_o \log(1 - p(\alpha'z_i)) \qquad (2.7)$$

where $f(.)$ is the marginal density of w_i and $\sum_+$ and $\sum_o$ refer respectively to summation over positive and zero recorded expenditures.

The model described by (2.5) and (2.6) has so far been set in a framework where there is only one commodity (i.e. a single item) for which a purchase either takes place or does not take place during the survey period. Our probability term p_i can however be thought of more generally as the probability of *at least* one purchase in the survey period of *at least* one of a number of items that collectively define the commodity group. The example we use in the empirical section is household clothing expenditure. If there are for example two items purchased with probability p_i^1 and p_i^2 by household i, then if these purchases were independent p_i would simply equal $p_i^1 + p_i^2 - p_i^1 p_i^2$. Similarly since for each item $E(e_i^j) = c_i^j$ we have $E(e_i) = E(c_i)$ and we know $E(e_i|e_i>0) = E(c_i)/p_i$ which underlies the stochastic consumption model (2.4).

The only problem that remains when considering a group of items is the problem of the aggregation of Engel curves across items. For aggregation purposes, it may be preferred to adopt a linear form for (2.1). However, apart from the complication brought about by a non-unitary Jacobian term having now rewritten (2.5) as

$$p_i e_i = \theta'x_i + w_i^* \qquad (2.8)$$

and (2.4) as

$$c_i = p_i e_i + v_i^* \qquad (2.9)$$

the parameters θ and p_i would be recovered in the exact manner described above.

3. Empirical results for a model of household expenditures on clothing

Clothing expenditures by households in a survey would seem one of the most appropriate commodity groupings for illustrating the procedures developed in

Section 2. The consumption of services from clothing by households must (by casual observations!) be positive. On the other hand even aggregating across *all* clothing expenditures in a household over a survey period some zero expenditures are still likely. For example, in the data we use which cover a sample of married working age couples from the UK Family Expenditure Survey, around 12% of the sample record zero expenditures over the two week survey period. A description of the distribution of expenditures for this commodity group and for two additional comparison groups (tobacco and services) is provided in the Data appendix.

Since the Engel curve, seen as an income-expenditure equation, can be thought of as a reduced form from a set of structural relationships describing all expenditure, savings and even labour market decisions of the household a variety of economic and demographic variables can be expected to enter the determination of consumption. The probability of purchase on the other hand may depend more directly on variables that determine the relative time and money costs of purchase.

There have been a number of previous attempts to model expenditures in the Family Expenditure Survey including those of Atkinson, Gomulka and Stern (1984), Deaton and Irish (1984) and Kay, Keen and Morris (1984) mentioned above which concern themselves with the occurrence of zero expenditures directly and those of Atkinson and Stern (1980), Blundell (1980), Blundell and Walker (1982) and Blundell and Ray (1984) which were more concerned with the appropriate specification of economic and demographic factors. From these studies it is apparent that family size and age variables should be carefully specified and that both income *and* hourly wage variables are likely to be important determinants of consumption. In addition to these we enter education and labour market variables that can be expected to influence the level of consumption and the degree of 'stocking-up' through future income expectations. However, as explained in the introductory comments to this paper we do not attempt to identify preference parameters directly in this study.

The maximum-likelihood parameter estimates for the consumption function developed in Section 2 are presented with standard errors in Table 1. The older child groups have significant coefficients both on their own and when interacted with the log income variable. Similarly, the hourly wage rate w_m enters significantly both on its own and when interacted with income. Finally the log income squared enters significantly. The overall properties of this model specification seem reasonable. The parameter estimates on significant variables were found to be robust on the inclusion of more general labour market and other locational variables. The heteroscedasticity test, which is a five degrees of freedom test, is again relatively acceptable.

The frequency-of-purchase probability whose estimated parameters are presented in Table 2 have equally acceptable properties. Perhaps even more important here are the diagnostics against stochastic misspecification which indicate an overall acceptance of the normality assumption. These tests are derived from the Lagrange Multiplier tests developed in Bera, Jarque and Lee (1984), Blundell and Meghir (1986a), Chesher and Irish (1984) and Gourieroux, Monfort, Renault and Trognon (1984). Rather surprisingly, the independence test (in Table 1) – that is the independence of the unobservables in the Engel curve and latest frequency-of-purchase variable model – indicates a degree of acceptance. Apart from some of the income terms in Table 2 the interesting features relate to the regional and employment dummies. Although male employment had no apparent effect on the level of consumption it affects purchase frequency quite dramatically with a partially offsetting effect for those households in the London region. It is perhaps worth noting at this stage that least-squares estimation of the *linear* Engel curve model using all observations would generate consistent parameter estimates under independence (see Keen 1985)).

In order to interpret these parameter estimates more fully we present some

Table 1. Parameter estimates and the independence test for the Engel curve.*

Variable	Parameter estimate	Standard error
Intercept	1.187	(2.172)
n_1	0.882	(0.837)
n_2	− 0.570	(0.901)
n_3	− 1.215	(0.440)
n_4	0.391	(0.471)
Age	− 0.027	(0.030)
Education	0.001	(0.022)
Male employed	0.064	(0.207)
London	0.197	(0.132)
log(income)	− 0.823	(0.547)
$\log^2$(income)	0.170	(0.042)
log(income) · n_1	− 0.185	(0.171)
log(income) · n_2	0.176	(0.187)
log(income) · n_3	0.254	(0.089)
log(income) · n_4	− 0.018	(0.093)
log(income) · Age	0.005	(0.008)
log(income) · log(wage)	− 0.690	(0.228)
log(wage)	3.183	(0.962)
$\log^2$(wage)	0.286	(1.527)
σ	1.302	(0.024)
LogL	− 3125.04 Heteroscedasticity (5) 10.84	
Independence (1)	1.974	

* Standard errors in parentheses.

indicative income and wage elasticities in Table 3. Clothing consumption turns out, as expected, to be a normal good. It becomes more of a necessity with the presence of young children while the presence of children between 5 and 11 years of age increases the elasticity toward unity. The wage elasticity on consumption is similarly positive. Rather interestingly, the frequency-of-purchase probability has a positive wage and income elasticity in general although the presence of children in the second age group tends to reverse this effect. The importance of interactions between income and demographics is clearly demonstrated in these elasticity measures.

4. Conclusions

This paper has been concerned with the maximum-likelihood estimation of Engel curve parameters using individual survey data. In particular, emphasis was placed on the infrequency-of-purchase model where zero expenditures are recorded during the survey period due to the overall 'durability' of the

Table 2. Parameter estimates for purchase probability and non-normality diagnostics.

Variable	Parameter estimate	Standard error
Quarter 1	−0.073	(0.215)
Quarter 2	−0.108	(0.120)
Quarter 3	0.093	(0.129)
Intercept	0.0254	(2.620)
D1	0.147	(0.320)
D2	0.847	(0.475)
D3	−0.141	(0.206)
Age	0.037	(0.035)
Education	−0.009	(0.025)
London	−0.398	(0.130)
Male Employed	0.589	(0.218)
log(income)	−0.201	(1.071)
log(income)^2	0.069	(0.119)
log(income) · n_1	0.023	(0.055)
log(income) · n_2	0.134	(0.089)
log(income) · n_3	0.041	(0.027)
log(income) · n_4	0.046	(0.013)
log(income) · Age	−0.008	(0.007)
log(income)log(wage)	−0.273	(0.291)
log(wage)	1.350	(1.207)
log(wage)^2	0.093	(0.2443)
Skewness (1) 3.017	Kurtosis (1) 2.705	Non-normality 3.082

* Standard errors in parentheses.

commodity in question. In this model, when purchases are made, expenditure tends to overestimate consumption. The expectation of expenditure conditional on purchasing overestimates consumption by a factor equal to the inverse of the purchase probability. This conditional expectation forms the basis for our estimation procedure which allows both the purchase probability and the consumption Engel curve to depend quite generally on a variety of economic and demographic characteristics.

Our illustration relates to the consumption and purchase of clothing by a sample of households in the UK Family Expenditure Survey. A loglinear Engel curve model is chosen although the theoretical concepts are developed to cover any form. The purchase probability is assumed to be a cumulative normal and to be independent of the unobservable components in the Engel curve relationship. These assumptions are tested using appropriate Lagrange Multiplier tests and found acceptable in our empirical example. The overall impact of income and other economic as well as demographic characteristics was found significant and plausible giving solid motivation for the further use and development of this Engel curve model.

5. Acknowledgement

We should like to thank Javad Emami for excellent research assistance, the Institute for Fiscal Studies and the Department of Employment for providing the data used in this study. Finance for this research was provided under the ESRC grant B0023 2060.

*Table 3. Some income and wage elasticities.**

(i) Income elasticities Family structure	Consumption	Purchase frequency
$n_1 = n_2 = n_3 = n_4 = 0$	0.617	0.003
$n_1 = 1\ n_2 = n_3 = n_4 = 0$	0.432	0.008
$n_1 = 0\ n_2 = 1\ n_3 = n_4 = 0$	0.793	-0.025
$n_1 = n_2 = 0\ n_3 = n_4 = 0$	0.871	0.012
$n_1 = n_2 = n_3 = 0\ n_4 = 1$	0.599	0.013
$n_1 = n_2 = 1\ n_3 = n_4 = 0$	0.607	-0.019
$n_1 = 1\ n_2 = 0\ n_3 = 1\ n_4 = 0$	0.685	0.016
(ii) Wage Elasticities		
Consumption	0.131	
Frequency of purchase	0.024	

* Evaluated at Age = 36, Education = 16, log(income) = 4.95 and log(wage) = 0.6346.

6. Data Appendix

The Data were drawn from the Family Expenditure Survey 1981. The following sample selection was applied
a) Occupation of workers.
   Clerical, Shop Assistants, Manual Workers (skilled, semi-skilled and unskilled)
   i.e. FES variable A210 with value, 4, 5, 6, 7, 8.
b) Age of adults
   Women with 16<Age<60
   Men with 16<Age<65
c) Two adult households with the two adults being a couple
   The Resulting Sample Size was 1809

Expenditure Frequencies (Annualised figures in £'s)
(i) *Clothing*

expenditure	cumulated percentage
0	11.9
0.01 and less than 10	14.2
10.01 and less than 20	16.9
20.01 and less than 50	22.7
50.01 and less than 100	29.0
100.01 and less than 150	35.2
150.01 and less than 300	51.8
300 and above	100.0

(ii) *Tobacco*

expenditure	cumulated percentage
0	39.1
0.01 and less than 10	39.3
10.01 and less than 20	39.6
20.01 and less than 50	41.4
50.01 and less than 100	45.3
100.01 and less than 150	50.3
150.01 and less than 300	67.3
300 and above	100.0

(iii) *Services*

expenditure	cumulated percentage
0	0.1
0.01 and less than 10	0.2
10.0 and less than 20	0.2

20.01 and less than 50	1.5
50.01 and less than 100	4.3
100.01 and less than 150	9.3
150.01 and less than 300	34.8
300 and above	100.0

Data Analysis

Variable	Mean	Standard deviation
Clothing Expenditure (annual £'s)	491.879	665.903
Total Expenditure (")	3692.150	1883.550
Children (age<2) (n_1)	0.230	0.497
Children (age 2–4) (n_2)	0.135	0.362
Children (age 5–10) (n_3)	0.485	0.763
Children (age>10) (n_4)	0.440	0.763
Male wage (hourly after tax)	2.085	0.980
Age (wife)	36.264	10.970
Education (at leaving education)	15.412	1.930
Male exmployment (zero-one dummy)	0.934	0.250
Household net income (log annual £'s)	4.946	0.530

Bibliography

Amemiya T. 1973. Regression analysis when the dependent variable is a truncated normal. *Econometrica*, 41: 1193–1205.

Atkinson AB and Stern NH. 1980. On the switch from direct to indirect taxation. *Journal of Public Economics*, 14: 195–224.

Atkinson AB, Gomulka J and Stern NH. 1984. *Household expenditure on tobacco 1970–1980*. London School of Economics ESRC Programme on Taxation, Incentives and the Distribution of Income, Discussion Paper 57.

Bera AK, Jarque CM and Lee LF. 1984. Testing for the normality assumption in limited dependent variable models. *International Economic Review*, 25: 563–578.

Blundell RW. 1980. Estimating continuous consumer equivalent scales in an expenditure model with labour supply. *European Economic Review*, 40: 145–157.

Blundell RW and Walker I. 1982. Modelling the joint determination of household labour supplies and commodity demands. *Economic Journal*, 42: 351–364.

Blundell RW and Ray R. 1984. Testing for linear Engel curves in a new flexible demand system. *Economic Journal*, Conference Papers.

Blundell RW and Meghir C. 1986a. Selection criteria for a microeconometric model of labour supply. *Journal of Applied Econometrics*, 1: 57–84.

Blundell RW and Meghir C. 1986b. Bivariate alternatives to the tobit model. Forthcoming in *Journal of Econometrics*.

Chesher A and Irish M. 1984. Residuals and diagnostics for probit, tobit and related models. *Economics discussion paper*, University of Bristol.

Cragg JG. 1971. Some statistical models for limited dependent variables with applications to the demand for durable goods. *Econometrica*, 39: 829–844.

Cramer JS. 1964. Efficient grouping, regression and correlation in Engel curve analysis. *Journal of the American Statistical Association*, 59: 656–668.

Cramer JS. 1971. *Empirical Econometrics*. Amsterdam: North-Holland Publishing Co.

Deaton AS and Muellbauer J. 1980. *Economics and consumer behaviour*. Cambridge: Cambridge University Press.

Deaton AS and Irish M. 1984. Statistical models for zero expenditures in household budgets. *Journal of Public Economics*, 23: 59–80.

Gourieroux C, Monfort A, Renault E and Trognon A. 1984. *Residus generalisés ou interpretations lineaires de l'econométrie non lineaire*. INSEE document de travail no. 8410, April.

Kay JA, Keen MJ and Morris CN. 1984. Estimating consumption from expenditure data. *Journal of Public Economics*, 23: 169–182.

Keen MJ. Zero expenditure and the estimation of Engel curves. Institute of Fiscal Studies, working paper, forthcoming in *Journal of Applied Econometrics*.

Pudney SE. 1985. *Frequency of purchase and Engel curve estimation*. Mimeo, London School of Economics, August.

Estimating and testing an almost ideal demand system

S. GEORGANTELIS, GARRY D.A. PHILLIPS and W. ZHANG

1. Introduction

Since the introduction of the Almost Ideal Demand System (AIDS) in the seminal paper by Deaton and Muellbauer (1980), few applications of their model have been reported. This may be due partly to the relatively complicated structure of the model and the associated estimation problems but it may also be due to the fact that the simplified versions of AIDS recommended by Deaton and Muellbauer have been used instead. More specifically, when it is possible to approximate a highly nonlinear price index (deflator) by a much simpler index such as the Stone (1953) index and hence reduce the level of computational difficulty to that of, say, the Rotterdam model, it is not surprising that researchers choose not to employ the AIDS model but use instead a model based on the simpler index.

In this paper we examine whether the use of the simplified version rather than the full AIDS model is justified and this is done in the context of the specification and estimation of a 10-equation demand system based on annual UK food data for the period 1950–83. This is a longer time period than that used by Deaton and Muellbauer (1980) who based their analysis on UK data for the period 1954–1974 and who used an 8-equation complete UK demand system. A feature of the extended data series is that it not only provides more observations and hence, more degrees of freedom in the estimation, but also it includes a period when prices have changed quite dramatically. Our choice of a system of demand equations for food rather than using a complete system of demand equations was made so as to reduce the problems associated with aggregation and omitted dynamic elements. Our results suggest that the full AIDS model can be estimated and used to test the theoretical restrictions of demand theory and that the results may be an improvement over those obtained by using simpler models. This tendency is particularly evident when, as in our study, the ideal price index is not well approximated by other common price indices.

R.D.H. Heijmans and H. Neudecker (eds.), The Practice of Econometrics. ISBN 90-247-3502-5.
© 1987, Martinus Nijhoff Publishers, Dordrecht. Printed in the Netherlands.

2. The AIDS model

The Almost Ideal Demand System (AIDS) assumes that consumer preferences fall within the PIGLOG (price-independent generalised logarithmic) class so that exact aggregation over consumers is possible. These preferences are represented via the cost function which defines the minimum expenditure necessary to attain a specific utility at given prices. With a suitable choice of cost function, see Deaton and Muellbauer (1980, p. 313), the demand functions can be derived cf. Diewert (1974), Shephard (1953, 1970). The AIDS demand functions may be written as

$$w_{it} = \alpha_i + \sum_{j=1}^{n} \gamma_{ij} \log p_{jt} + \beta_i \log\left(\frac{Y_t}{P_t}\right), \quad i=1,\ldots,n, \quad t=1,\ldots,T, \quad (1)$$

with

$$\log P_t = \alpha_0 + \sum_{k=1}^{n} \alpha_k \log p_{kt} + \frac{1}{2} \sum_{j=1}^{n} \sum_{k=1}^{n} \gamma_{kj} \log p_{kt} \log p_{jt}. \quad (2)$$

Here w_i is the share of aggregate expenditure on good i, the p_j are the prices of goods, Y is total expenditure and P is a price index. The adding-up property leads to the following restrictions:

$$\sum_{i=1}^{n} \alpha_i = 1, \sum_{i=1}^{n} \beta_i = 0, \sum_{i=1}^{n} \gamma_{ij} = 0, j=1,\ldots,n. \quad (3)$$

The AIDS model gives an arbitrary first-order approximation to any demand system, such as the Rotterdam model, and it satisfies the axioms of choice exactly. Moreover, it enables perfect aggregation over consumers without requiring parallel linear Engel curves. The possession of these properties makes the model an especially good vehicle for testing the homogeneity restrictions

$$\sum_{j=1}^{n} \gamma_{ij} = 0 \quad (4)$$

and the symmetry restrictions

$$\gamma_{ij} = \gamma_{ji}. \quad (5)$$

As with any complete demand system, one can estimate $n-1$ of the equations and then the parameter estimates for the remaining equations may be obtained using the constraints in (3).

The homogeneity conditions in (4) impose $n-1$ restrictions while the symmetry conditions in (5) impose a further $\frac{n(n-1)}{2}$.

3. A comparison of the AIDS and Rotterdam models

The substitution of (2) into (1) yields:

$$w_{it} = (\alpha_i - \beta_i \alpha_0) + \sum_{j=1}^{n} \gamma_{ij} \log p_{jt} + \beta_i \{\log Y_t - \sum_{k=1}^{n} \alpha_k \log p_{kt} -$$

$$-\frac{1}{2} \sum_{j=1}^{n} \sum_{k=1}^{n} \gamma_{kj} \log p_{kt} \log p_{jt}\} \qquad i = 1, \ldots, n. \qquad (6)$$

These equations automatically satisfy the adding-up restrictions in (3) and they represent a non-linear system of demand equations.[1] All the parameters in (6) are identified and can be estimated.[2] Deaton and Muellbauer (1980) note that the estimation of (6) would be much simpler if P were known; in fact the situation would then be comparable to that when using the Rotterdam model. They suggest that when prices are highly collinear then, for some constant φ,

$$P_t \simeq \varphi P_t^* \qquad (7)$$

where

$$\log P_t^* = \sum_{k=1}^{n} w_{kt} \log p_{kt} \qquad (8)$$

is the well-known Stone (1953) index. In such a case, the AIDS demand functions are well approximated by setting P_t to P_t^* in (1) to yield

$$w_{it} = \alpha_i^* + \sum_{j=1}^{n} \gamma_{ij} \log p_{jt} + \beta_i \log\left(\frac{Y_t}{P_t^*}\right). \qquad (9)$$

The Rotterdam model demand functions are given by

$$w_{it} \triangle \log q_{it} = \sum_{j=1}^{n} c_{ij} \triangle \log p_{jt} +$$

$$b_i \{\triangle \log Y_t - \sum_{k=1}^{n} w_{kt} \triangle \log p_{kt}\}; \qquad i = 1, \ldots, n. \qquad (10)$$

where $\triangle$ represents a first difference and the q_{it} are the quantities purchased.

If the equation in (6) is expressed in first difference form we have

---

[1] This means that the variance-covariance matrix of the disturbances is singular.
[2] Deaton and Muellbauer (1980 p. 316) maintain that 'in many examples the practical identification of α_0 is likely to be problematical . . . this can be overcome by assigning a value to α_0 a priori. Since the parameter can be interpreted as the outlay required for a minimal standard of living when prices are unity choosing a plausible value is not difficult.'

$$\Delta w_{it} = \sum_{j=1}^{n} \gamma_{ij} \Delta \log p_{jt} + \beta_i \{\Delta \log Y_t - \sum_{k=1}^{n} \alpha_k \Delta \log p_{kt} - \frac{1}{2} \sum_{j=1}^{n} \sum_{k=1}^{n} \gamma_{kj} \times \Delta(\log p_{kt} \log p_{jt})\} \qquad (11)$$

which may be compared with (10) and it is seen that except for the different price indices the right-hand sides are the same.[3] Deaton and Muellbauer (1980) note that if we use the equality

$$\Delta \log w_i = \Delta \log p_i + \Delta \log q_i - \Delta \log Y$$

and replace $\Delta \log q_{it}$ with $\Delta \log w_{it}$ in (10), then, upon rearrangement, the first difference form of the AIDS model is generated.

The main advantage the AIDS model has over the Rotterdam model is that the price index P may be directly estimated from (2) and that the AIDS equation (6) is derived from an explicit cost function. The former property has important implications for the calculation of cost-of-living indices which are consistent with demand theory considerations while the latter makes the AIDS model especially attractive for use in testing the validity of the restrictions of neoclassical demand theory, see Deaton and Muellbauer (1980, p. 315).

4. Data and estimation

The data to be analysed concern the allocation of UK food expenditure among 10 food categories under the assumption that utility is weakly separable in food versus other items. This assumption is required so that the constrained maximization problem associated with the food branch of the utility tree may be considered separately from the first-stage allocation of income to food and non-food expenditure. The food categories used were:
1. Bread and cereals
2. Meat and bacon
3. Fish
4. Oils and fats
5. Sugar, preserves and confectionery[4]

---

[3] Deaton and Muellbauer note that when prices are nearly collinear estimation of (11) is difficult so that the practical approach is to replace $\Delta \log P = \sum \alpha_k \Delta \log p_{kt} + \frac{1}{2} \sum_{k=1}^{n} \sum_{j=1}^{n} \gamma_{kj} \Delta (\log p_{kt} \log p_{jt})$ by some index, e.g. $\Delta P^* = \Delta(\sum_{k=1}^{n} w_k \log p_{kt})$ or the Rotterdam deflator $\sum_{k=1}^{n} w_{kt} \Delta \log p_{kt}$. These suggestions were followed when the first difference form of the AIDS model was estimated.

6. Dairy products
7. Fruit
8. Potatoes and vegetables[5]
9. Beverages[6]
10. Other manufactured food.

Data covering expenditure at nominal and constant prices for the period 1950–1983 were taken from successive CSO United Kingdom National Accounts Bluebooks from which data on budget shares, prices and other relevant variables were constructed.

The AIDS model was used in estimating the demand functions for the food categories and, for comparative purposes, it was decided to estimate the Rotterdam model based on the same data.

It is important to distinguish the approach followed in this paper from that used by Deaton and Muellbauer (1980). They first estimated the simpler model (9) by unrestricted OLS after setting $\log P^* = \sum_{k=1}^{n} w_k \log p_k$. The equations were then re-estimated by OLS, again using P^*, in order to test the homogeneity restrictions (4). Next both the homogeneity and symmetry restrictions (5) were imposed and the full AIDS model (6) was estimated after imposing the restriction $\Omega = \sigma^2(I - \underline{i}\underline{i}')$ on Ω, the covariance matrix of the vector of disturbances across the equations.[7] Here $\underline{i}$ is a constant $n \times 1$ vector with each component equal and given by $\sqrt{\frac{1}{n}}$, n is the number of equations to be estimated and I is an $n \times n$ identity matrix. Using the restricted parameter estimates they constructed an estimate of P according to (2) and this index was compared with P^*. The effect on estimation of using P rather than P^* was examined by re-estimating the unrestricted and homogenous forms of (9), replacing P^* with the estimate of P.[8] Finally, symmetry was tested in the context of homogeneity using the same model and parameter estimates were obtained which satisfied both homogeneity and symmetry restrictions.[9]

In our estimation we used the Iterative Seemingly Unrelated Regression

---

[5] In the 1984 Bluebook this group was further disaggregated.

[6] This group is also disaggregated into coffee, tea and cocoa, and soft drinks.

[7] This was done because there were too few degrees of freedom to obtain the maximum-likelihood estimate of Ω. The approach used reduces to non-linear OLS subject to the homogeneity and symmetry restrictions.

[8] Calculating P from homogeneity and symmetry constrained estimates and then using it as a deflator in both homogeneity-constrained and unconstrained versions of the model implies that any index P' calculated from the unconstrained model should be close to P. In our case P was estimated directly.

[9] The estimate of P was used in the AIDS demand functions when testing the homogeneity and symmetry restrictions. We also test symmetry in the context of homogeneity notwithstanding the objections to this, see Deaton and Muellbauer p. 318.

Table 1. AIDS model in levels: Homogeneity and Symmetry Constrained Parameter Estimates.*

Food commodity i	α_i	β_i	γ_{i1}	γ_{i2}	γ_{i3}	γ_{i4}	γ_{i5}	γ_{i6}	γ_{i7}	γ_{i8}	γ_{i9}	γ_{i10}	R^2	DW
Bread and cereals	0.411 (13.5)	−0.186 (−9.2)	0.001 (0.1)										0.941	1.21
Meat and bacon	−0.192 (−4.8)	0.316 (12.1)	−0.010 (−0.6)	−0.004 (0.0)									0.924	2.20
Fish	0.104 (5.5)	−0.050 (−4.0)	0.021 (3.2)	0.002 (0.2)	0.010 (1.9)								0.320	1.37
Oils and fats	0.053 (4.4)	−0.008 (−1.0)	−0.019 (−4.7)	−0.010 (−1.5)	−0.005 (−1.9)	0.037 (13.0)							0.966	1.34
Sugar, preserves and confectionery	0.206 (6.6)	−0.068 (−3.3)	−0.041 (−4.3)	0.063 (3.8)	−0.036 (−5.9)	0.002 (0.5)	0.061 (4.6)						0.520	1.90
Dairy products	0.294 (10.7)	−0.107 (−6.0)	−0.009 (−1.2)	0.020 (1.4)	−0.002 (−0.4)	−0.013 (−3.5)	−0.029 (−3.3)	0.078 (6.7)					0.835	0.79
Fruits	0.012 (0.6)	0.030 (2.3)	0.023 (4.3)	−0.034 (−3.3)	−0.007 (−1.8)	0.012 (4.3)	−0.015 (−2.7)	0.003 (0.5)	0.061 (12.0)				0.803	2.40
Potatoes and vegetables	0.121 (6.2)	0.000 (0.0)	0.014 (2.6)	−0.012 (−1.1)	0.009 (2.7)	−0.006 (−2.3)	−0.016 (−2.8)	−0.019 (−3.5)	−0.008 (−2.4)	0.066 (13.4)			0.840	1.26
Beverages	−0.081 (−4.1)	0.096 (7.3)	−0.005 (−0.7)	−0.044 (−4.0)	0.022 (5.1)	0.008 (2.3)	0.034 (5.4)	−0.004 (−0.5)	−0.015 (−3.1)	−0.030 (−8.0)	0.018 (2.8)		0.814	1.06
Other manufactured food	0.070	−0.023	0.024	0.028	−0.015	−0.007	−0.023	−0.026	−0.021	0.003	0.015	0.021		

* (t-values in parentheses).

$\alpha_0 = -4.891$.

(ITSUR) estimation technique which yields estimates which are asymptotically equivalent to ML estimates. The full AIDS model (6) was first estimated directly (without fixing α_0) in its unconstrained form and then re-estimated subject to the homogeneity restrictions. Tests of homogeneity and symmetry were carried out directly using (6) and finally, homogeneity and symmetry restricted estimates were obtained. The index P was calculated from these estimates and compared with the index P*. In obtaining these estimates we did not restrict Ω.

A somewhat similar analysis was carried out for the AIDS model in first differences (11). However, when estimating the AIDS model in first differences we used the P index calculated for the levels case.

The results obtained were compared with those based upon a first difference form of (9) in which P* or a Rotterdam-type deflator was used. In addition, the Rotterdam model (10) itself was included as a further basis for comparison.[10]

5. Results: AIDS model in levels

Model (6) was estimated subject to the adding-up, homogeneity and symmetry restrictions by directly imposing these restrictions on the original AIDS model. Even after the restrictions were imposed there remained the problem of estimating α_0, see Deaton and Muellbauer (1980, p. 316). Instead of choosing a plausible value for α_0 which was the approach followed by Deaton and Muellbauer (1980), we imposed a plausible interval within which α_0 was constrained to lie.[11]

Table 1 presents the homogeneity and symmetry constrained estimated parameters of the AIDS model. The estimated β_i suggest that Bread and Cereals, Fish, Oils and Fats, Sugar and Preserves, Dairy Products and Other Manufactured Goods are all necessities whereas Meat and Bacon, Fruits and Beverages can be classified as luxuries. A large number of the γ_{ij} are significantly different from zero. The impact of the sign and magnitude of the γ_{ij} can be assessed in Table 2 where estimated own-price and income elasticities are presented, see the Appendix for method of calculation.

Next we consider the effect of approximating (2) by (8) i.e. estimating the simpler version of the AIDS model[12] replacing P_t with P_t^*. The associated own-

[10] We introduced an intercept term into (11) and in the Rotterdam model, to allow for possible time trends.
[11] This interval was formed by the first observation for total real expenditure and its 1975 (base year) value. Estimation was carried out in two stages. Firstly the model was estimated subject to the interval constraint; then α_0 was fixed at its estimated value and the remaining parameter estimates were obtained. By imposing constraints i.e. homogeneity and/or symmetry different estimates of α_0 were obtained. Deaton and Muellbauer fixed α_0.
[12] When the full AIDS model in (6) is estimated directly we shall say that P is used as the deflator, see (1) and (2).

price and income elasticity estimates are presented in the final two columns of Table 2.

Notice that there are obvious differences between the two estimates depending upon the deflator used; this is most noticeable for the own-price elasticity estimates.[13] These differences are explained by the fact that the computed ideal price index in (2) often diverges from the simpler Stone index in (8) as can be seen from Table 3.

Finally Tables 4 and 5 present the results of testing the homogeneity and the symmetry restrictions using the AIDS model. Table 4 indicates that for both the full and simple versions of the model the homogeneity restrictions are rejected based on asymptotic Wald (W), Likelihood-Ratio (LR) and Lagrange-Multiplier (LM) tests. However, when the critical values are corrected using Edgeworth expansions, see Rothenberg (1977), the modified LM test, LM**, fails to reject homogeneity at the 1% level of significance.

Table 5 presents the results of testing the symmetry restrictions in the context of homogeneity. These restrictions are rejected for both models even at the 1% level of significance. This is so for all tests including the most conservative of them, i.e. the LM test.

The above results are in accordance with other findings; they also indicate

Table 2. Total expenditure and own-price elasticities using P and P* deflators.

Food commodity i	w_i	P		P*	
		e_i	e_{ii}	e_i	e_{ii}
Bread and cereals	0.137	−0.355	−0.432	−0.400	−0.576
Meat and bacon	0.267	2.182	−0.775	2.176	−0.473
Fish	0.033	−0.538	−0.514	−0.437	−0.559
Oils and fats	0.047	0.837	−0.201	0.813	−0.212
Sugar preserves and confectionery	0.099	0.315	−0.250	0.416	−0.323
Dairy products	0.153	0.302	−0.270	0.293	−0.398
Fruits	0.060	1.496	0.004	1.562	0.088
Potatoes and vegetables	0.111	1.000	−0.406	0.974	−0.397
Beverages	0.061	2.564	−0.577	2.470	−0.481
Other manufactured food	0.032	0.247	−0.292	0.295	−0.313

Note: w_i, e_i and e_{ii} denote budgetshares, income elasticities and own-price elasticities respectively. P is the directly estimated almost ideal price index. P* is Stone's index.

[13] Deaton and Muellbauer's estimates were much closer presumably because a shorter time period was used i.e. 1954–74, during which prices were relatively stable. Here own-price elasticities differ more than income elasticities although all the corresponding estimates show consistency with respect to sign and size. Bread and Fish appear as inferior goods with Fruits exhibiting an own price elasticity with the wrong sign but very close to zero.

Table 3. Comparison of price indices.

Year	P	P*	Year	P	P*	Year	P	P*
1950	0.269	0.210	1961	0.376	0.352	1972	0.616	0.592
1951	0.287	0.238	1962	0.392	0.364	1973	0.642	0.679
1952	0.310	0.268	1963	0.405	0.372	1974	0.786	0.793
1953	0.326	0.282	1964	0.409	0.382	1975	1.000	1.000
1954	0.322	0.295	1965	0.423	0.397	1976	1.144	1.199
1955	0.323	0.319	1966	0.435	0.412	1977	1.349	1.441
1956	0.349	0.334	1967	0.445	0.421	1978	1.481	1.590
1957	0.364	0.341	1968	0.470	0.441	1979	1.690	1.806
1958	0.366	0.346	1969	0.495	0.467	1980	1.927	2.074
1959	0.366	0.348	1970	0.525	0.495	1981	2.061	2.246
1960	0.366	0.346	1971	0.580	0.551	1982	2.187	2.418

Note: P is the directly estimated almost ideal price index.
P* is Stone's index.

that the choice of deflator makes little difference to the tests.[14] When homogeneity was imposed (using either P or P*) there was a sharp drop in the value of the DW statistic. Similar results have been reported by other researchers, e.g. Barten (1969) and Deaton and Muellbauer (1980), and possible reasons for this phenomenon include the omission of relevant variables, particularly trending variables.

When symmetry was imposed, in addition to homogeneity, there was an even larger drop in the DW statistic. Again, possible reasons for the rejection of symmetry include the neglect of habit-formation factors and dynamic elements in the specification of the AIDS model. Clearly testing for symmetry

Table 4. Homogeneity tests using P and P* deflators.

Deflator	Asymptotic test			χ^2 cvs		Modified tests			Edgeworth χ^2 cvs. LM**	
	W	LR	LM	0.05	0.01	W**	LR**	LM**	0.05	0.01
P	190.85	64.09	28.67	16.92	21.67	123.21	42.29	19.32**	16.59	20.99
P*	195.76	64.96	28.97			126.26	42.86	19.53**		

Note: P is the directly estimated almost ideal price index,
P* is Stone's index.
** denotes not significant at the 1% level.

[14] Deaton and Muellbauer found no practical differences either.

in the context of rejected homogeneity constraints makes a clear interpretation of the test outcomes difficult. Again it is seen that the choice of deflator has little effect on the tests. However, in the next section we shall examine several AIDS models in first-difference form and it will be seen that the choice of deflator in such models is much more important.

6. Results: a comparison of AIDS models in first differences and the Rotterdam model

In this section we examine the estimation and testing of the AIDS model in first difference form, see (11), and we compare the results with those obtained for the Rotterdam model. The form of the AIDS model estimated[15] included an intercept to account, to some degree, for the influence of time trending variables not explicitly included in the specification. The model was estimated using the various deflators suggested by Deaton and Muellbauer (1980), commencing with the calculated ideal index[16], $\triangle P$, followed by Stone's index, $\triangle P^*$, and, finally, using the Rotterdam deflator $\sum w_k \triangle \log p_k$. Our aim was to examine the effects on the estimated elasticities and the homogeneity and symmetry tests of using the different deflators and, in addition, to compare these results with those for the Rotterdam model. The result of testing the homogeneity restrictions are given in Table 6.

Note that homogeneity for the AIDS model is not rejected using the unmodified LM test based on any of the three deflators but the non-rejection is strongest for the $\triangle P$ deflator. Interestingly, homogeneity for the Rotterdam model is rejected using all three tests at either level of significance. When Rothenberg-type corrections were introduced the version of the AIDS model using the $\triangle P$ deflator again showed the strongest nonrejection of homogeneity while the restrictions were again rejected for the Rotterdam model based

Table 5. Symmetry tests using P and P* deflators (assuming homogeneity).

Deflator	W	LR	LM	χ^2 critical values	
				0.05	0.01
P	179.94	115.79	75.34	55.76	63.69
P*	169.65	107.69	75.17		

[15] We used the first difference form of (1) or (9).
[16] Note that here we used the P index calculated from the *levels* model. This approach was also used by Deaton and Muellbauer.

Table 6. Tests of homogeneity: the AIDS and Rotterdam models.

Model	Asymptotic test statistics			χ^2 cvs		Modified test statistics			Edgeworth χ^2 cvs					
	W	LR	LM	0.05	0.01	W**	LR**	LM**	W**		LR**		LM**	
									0.05	0.01	0.05	0.01	0.05	0.01
ΔP	33.81	23.28	16.71*	16.92	21.67	21.52**	17.28**	15.19*	27.36	37.49	20.55	26.31	13.73	15.13
ΔP^*	57.66	33.35	20.99**			36.94**	24.76**	19.08						
$\sum w_k \Delta \log p_k$	37.89	25.23	17.64**			24.11**	18.73**	16.04						
Rotterdam	87.81	42.83	21.99			55.88	31.80	21.81						

* Denotes not significant at 5% level.
** Denotes not significant at 1% level.

on the modified test statistics. Notice that despite the similarity between the AIDS model which uses the deflator $\sum w_k \triangle \log p_k$ and the Rotterdam model, somewhat contradictory results are obtained for the two models. This indicates that if this version of the AIDS model is approximated by the Rotterdam model then misleading results may be obtained. It is also clear that whereas replacing P with P* in the levels version of the AIDS model had little effect on the homogeneity tests, the same cannot be said for the first-difference version where the effect is seen to be substantial.

Table 7 presents the results for testing the symmetry restrictions in the context of homogeneity.

It is seen in Table 7 that although the W and LR tests reject symmetry for all models, the LM test does not do so for the $\triangle P$ and $\sum w_k \triangle \log p_k$ versions of the AIDS model at either significance level. The $\triangle P^*$ version is not rejected at the 1% level while symmetry is always rejected for the Rotterdam model. Hence the choice of deflator is seen to be important for the symmetry tests also.

Table 8 presents estimated income and own-price elasticities for all four models. The estimated income elasticities show a consistency of sign across all models with only Bread and Cereals emerging as an inferior good. As for the size of the estimated income elasticities, Fish shows the largest variation although there is also considerable variation for Potatoes and Vegetables. On a priori grounds the estimates based on the $\triangle P$ or $\triangle P^*$ versions of the AIDS model seem the most reasonable.

The own-price elasticity estimates, with one exception, have the expected negative sign and fall within the interval $(-1, 0)$. The only exception occurs with the Rotterdam model where Beverages exhibit a positive own-price elasticity. Notice that all own-price elasticities for the Rotterdam model are uniformly lower in absolute value than the corresponding estimates for any of the versions of the AIDS model. For the AIDS models, the only commodity

Table 7. Tests of symmetry: the AIDS and Rotterdam models.

Model	W	LR	LM	χ^2 critical values	
				0.05	0.01
$\triangle P$	92.50	69.55	54.36*		
$\triangle P^*$	109.28	76.79	57.62**		
				55.76	63.69
$\sum w_k \triangle \log p_k$	104.78	73.99	55.59*		
Rotterdam	250.87	113.99	69.24		

* Denotes not significant at 5% level.
** Denotes not significant at 1% level.

groups for which there is much variation in the elasticities are Bread and Cereals and Meat and Bacon.

7. Conclusions

In this paper it has been shown that the full AIDS model can be used to explain the UK demand for Food, disaggregated into 10 constituent commodity groups. When estimated in levels a test of the homogeneity restrictions based on the LM test statistic did not lead to rejection either for the full AIDS model or for a simplified version which used Stone's price index. On the other hand, the symmetry restrictions were rejected in all three types of test employed. When the AIDS model is estimated in first differences neither homogeneity nor symmetry were rejected by the LM test and the test failed to reject homogeneity at a higher level of significance than in the levels case. In fact, even the Wald test failed to reject homogeneity. In this case the choice of price deflator proved to be of considerable importance in testing the neoclassical restrictions. The results using the AIDS model were superior to those based on the Stone price index, a result somewhat at variance with the empirical findings of Deaton and Muellbauer (1980). A Rotterdam-type deflator led to results which were closer to those for the full AIDS model but on the basis of the overall results there was no evidence to suggest that the Rotterdam model could replace the AIDS model. The estimated elasticities also indicated that the AIDS model was to be preferred.

Table 8. Total expenditure and own-price elasticities, using different price index deflators.

Food commodity i	$\triangle P$		$\triangle P^*$		$\sum w_k \triangle \log p_k$		Rotterdam	
	e_i	e_{ii}	e_i	e_{ii}	e_i	e_{ii}	e_i	e_{ii}
Bread and cereals	−0.225	−0.840	−0.074	−0.280	−0.469	−0.423	−0.515	−0.096
Meat and bacon	1.640	−0.948	1.422	−0.635	1.724	−0.645	1.504	−0.346
Fish	0.924	−0.736	0.393	−0.738	−0.003	−0.781	0.327	−0.588
Oils and fats	1.728	−0.250	1.325	−0.248	1.692	−0.262	1.725	−0.210
Sugar, preserves and confectionery	0.719	−0.544	1.060	−0.474	1.259	−0.504	1.524	−0.436
Dairy products	0.589	−0.280	0.936	−0.245	0.712	−0.216	0.871	−0.153
Fruits	2.352	−0.118	1.983	−0.102	2.487	−0.134	2.449	−0.032
Potatoes and vegetables	0.649	−0.274	0.704	−0.262	0.276	−0.281	0.270	−0.151
Beverages	1.542	−0.150	1.335	−0.217	1.638	−0.126	1.786	0.101
Other manufactured food	0.358	−0.279	0.863	−0.351	0.310	−0.293	0.188	−0.017

Note: The P index was calculated from the levels-model estimates.

A possible reason for the better performance of the first difference models compared with the models specified in levels could turn on the introduction of a constant term in the first difference equations. In effect, a time trend has been introduced and this may mitigate the effects of omitted variables.

We may conclude that on the basis of this study it is preferable to use the full AIDS model either in levels or in first differences. The model may be estimated directly in levels but when estimating the first difference form it is appropriate to use the almost ideal price index computed from estimation in levels. Other price deflators gave mixed results and the Rotterdam model did not appear to be an adequate approximation to the first difference version of the AIDS model which emerges here as the most appropriate vehicle for testing theoretical restrictions and obtaining elasticity estimates.

8. Appendix

In this appendix we present the formulae for calculating elasticities based on the AIDS-model parameter estimates. Omitting the constant term we can write the AIDS model as

$$w_i = \frac{p_i q_i}{Y} = \sum_{j=1}^{n} \gamma_{ij} \log p_j - \frac{1}{2} \beta_i \sum_{j=1}^{n} \sum_{k=1}^{n} \gamma_{kj} \log p_k \log p_j - \beta_i \sum_{k=1}^{n} \alpha_k \log p_k$$

or

$$q_i = \frac{Y}{P_i} \left\{ \sum_{j=1}^{n} \gamma_{ij} \log p_j - \frac{1}{2} \beta_i \sum_{j=1}^{n} \sum_{k=1}^{n} \gamma_{kj} \log p_k \log p_j - \beta_i \sum_{k=1}^{n} \alpha_k \log p_k \right\} \quad (A1)$$

Differentiating (A1) w.r.t. p_j and multiplying by $\frac{p_j}{q_i}$ yields

$$e_{ij} = \frac{1}{w_i} (\gamma_{ij} - e_i \sum_{k=1}^{n} \gamma_{jk} \log p_k - \beta_i \alpha_i). \quad (A2)$$

Similarly we may obtain

$$e_{ii} = \frac{1}{w_i} (\gamma_{ii} - \beta_i \sum_{k=1}^{n} \gamma_{ki} \log p_k - \beta_i \alpha_i) - 1 \quad (A3)$$

and

$$e_i = 1 + \frac{\beta_i}{w_i} \tag{A4}$$

Note that (A2), (A3) and (A4) yield the cross, own and income elasticities for the AIDS model in levels.

Bibliography

Barten AP. 1969. Maximum likelihood estimation of a complete system of demand equations. European Economic Review 1: 7–73.

Deaton A, Muellbauer J. 1980. An almost ideal demand system. The American Economic Review 70, 3: 312–326.

Diewert WE. 1974. Applications of duality theory, in: Frontiers of quantitative Economics, 2, ch. 3, edited by M.D. Intriligator and D.A. Kendrick. Amsterdam: North-Holland Publishing Co./ New York: American Elsevier Publishing Co.

Rothenberg TJ. 1977. Edgeworth expansions for some test statistics in multivariate regression, unpublished paper. University of California, Berkeley.

Shephard RW. 1953. Cost and production functions. Princeton: Princeton University Press.

Shephard RW. 1970. Theory of cost and production functions. Princeton: Princeton University Press.

Stone JRN. 1953. The measurement of consumer's expenditure and behaviour in the United Kingdom, 1920–1938, 1. Cambridge: Cambridge University Press.

Job separations and job matching

TONY LANCASTER, GUIDO IMBENS and PETER DOLTON

Preface

'I believe that the advance of applied economics is better served by the patient study of elementary economic phenomena than by the pursuit of larger theoretical structures, and we have set out to illustrate this view by the study of a very simple phenomenon indeed. If the results are few and offer no surprise, this is only as it should be.'

This was the point of view expressed by Cramer in 1961 near the conclusion of his elegant monograph on 'The Ownership of Major Consumer Durables'. It is arguable that job separation is an elementary, though far from simple, economic phenomenon – and we have tried to be patient.

1. Introduction

This paper reports a study of the probabilities of job separation of a sample of male University graduates in their first full-time jobs. We shall denote by $\theta(t)$ the probability of separation of an employee who has been employed for t months. A statisfactory econometric analysis requires a model for the function $\theta(t)$. The separation event arises out of the joint decision making of both employer and employee and to model it requires analysis of the behaviour of both parties. The difficulty of this problem perhaps accounts for the scarcity in the literature of models that can form the basis of a rigorous econometric analysis of tenure data. The principal exception to this rule is the job-matching model due to Jovanovic (1979) in which separation is viewed as an outcome of an optimal implicit contract to separate whenever the match appears, to both parties, to be inadequately productive. The matching model does take account of the jointness of the separation decision but appears to be some distance away from being, at least *a priori*, a fully satisfactory account. For example it neglects entirely the phenomenon of on-the-job search for alternative employment.[1]

The strategy taken in this paper is therefore not to claim to specify a complete econometric model for θ but rather to examine the data in the light of the matching model. We hope to see to what extent the data appear to be consistent with the matching model and to what extent there exist features of the data not clearly explicable in terms of the model. In the next section we shall briefly describe the matching model and in the following one we shall describe the data.

2. Job matching – an overview

A pairing of employee and employer is a match. The rate of output produced by the match per unit time period is the sum of three components. The first is a function g(x) of exogenous characteristics of the employee – for example his levels of education and training. This quantity g is known to both employee and employer and the same across all matches in an occupation. The second is a constant μ, unknown to either, and specific to the match. There exists a distribution of μ, of mean normalised to zero, across all possible matches and the value of μ for any particular match is a random drawing from this distribution. The third is a purely random component, ε_t, varying over time and distributed independently of x and μ.

Observation of the increments of output once a match has been made provides information about μ, so that at each subsequent date there exists a (posterior) distribution for the value of μ for that match given the output that it has generated to date. The mean of this distribution is m, which is a function of total output produced since the match began and the time that has elapsed. m is the expected value of μ given the output that the match has generated. The value of m when the match begins is the (prior) mean of μ, zero.

There will be values of m sufficiently low that the match will seem to be particularly unproductive and it will be reasonable for the employee to quit and to form another match. In this new match his *initial* expected productivity is once again g(x) since the prior expectation of μ on any match is zero. The value of m below which a quit will occur will depend on the date, since the precision of the posterior distribution of μ (whose mean is m) will depend on the date. A low value of m early on in the match will be discounted because little information about μ has yet been accumulated and its value is, as yet, quite uncertain. The same value of m later in the match, when μ is known

---

[1] Jovanovic (1984b) combines a matching model with optimal job search but empirical implementation of the model looks far from easy and we have not attempted it.

rather accurately, could well lead to a quit. Thus we can imagine a function of time, $\lambda(t)$, the barrier function, starting initially at a value some way below zero, such that a quit occurs at the first data at which, if it happens at all, the value of m falls below the value of the function λ.

We have remarked that a reasonable rule is to quit when and if m falls below λ. In fact Jovanovic argues that this is the form taken by an optimal rule, specifically an optimal implicit contract between employer and employee – Jovanovic (1979). Moreover under risk neutrality it is optimal for the contract to specify that the wage to be paid at each moment of time is equal to the expected marginal productivity (per unit time period) of the match and this is equal to $g(x) + m$. Note that we can therefore identify $g(x)$ as the starting wage w_0.

We can already see some implications of this model for tenure data. The separation process depends upon a process of learning about the match specific μ from the evidence provided by the match output or rather by the excess or shortfall of output per unit time over $g(x)$. This learning process does not depend on the value of $g(x)$ but depends only upon (a) prior beliefs about μ, and (b) the properties of ε_t, neither of which involve $g(x)$. On the other hand the separation process also depends upon the position of the barrier function λ. In the strict version of the model the barrier function does not involve $g(x)$ either. Thus we have the implication that job tenure does *not* depend upon the starting wage. This is testable with our data.

A more general version of the model would allow for the barrier to depend upon the starting wage. There are two reasons for this. The first, noted by Jovanovic, is that if a period of unemployment intervenes between jobs the opportunity cost of that unemployment will be the starting wage in a new job, $g(x)$. Unemployment will be more expensive the larger is g and thus we should expect to find the quit probability smaller the larger the starting wage. A second reason is that legal requirements for redundancy payments may make it more expensive to fire high-wage people. Since the wage depends upon $g(x)$ this would again lead to an inverse association between starting wage and separation probability.

A second implication of the model is that the separation probability should be initially very small. This is because little or no learning has taken place and there will be insufficient evidence on which to base a rational separation decision. On the other hand eventually μ is known with certainty, learning has ceased, and if a separation has not taken place by then it never will. Thus the separation probability should be initially zero and eventually fall to zero. The model predicts a *non-monotone* hazard function. Moreover in the strict version of the model in which the position of the threshold barrier, $\lambda(t)$, does not depend upon the measured component of productivity, g, then the separation probabilities do not, either. When employees are paid their perceived marginal product, so that g(x) is the starting wage, w_o, then separation probabilities are independent of the starting wage. This is an implication we shall test.

Furthermore, if employees are always paid their perceived marginal product the average wage paid to survivors must rise with tenure as those whose matches are poor, and who are correspondingly ill-paid, separate. This selection effect will ultimately cease when the match quality of survivors becomes known with certainty, and when the separation probability has fallen to zero the survivors' mean wage will cease to grow. Thus the wages of survivors should be an increasing, concave, function of tenure. Note that in this model the accumulation of human capital – apart from that capital represented by knowledge of the match quality – plays no role. We shall report the behaviour of the wages of survivors as a function of their tenure.

3. Data

The observations arise out of a survey of British University 1970 graduates conducted in 1977.[2] The sample used in this study consisted of men who satisfied the following criteria; have had at least one full-time job for at least one month, are not self-employed, were under 30 on graduation, reported their starting wage, and worked in a first job in private industry or commerce. For the persons in this sample we know the starting salary in the first job, date of starting and finishing first job and current (1977) salary. In addition occupation and sector of employment are known for that first job. The sample size was 1482.

4. Separation probabilities

4.1

We first consider all 1482 men and estimate a model in which the probability of separation is piecewise constant

$$\theta(t) = \theta_j \quad t_{j-1} < t \leq t_j; \; t_o = 0; \; t_k = \infty.$$

Of the 1482, 1057 had left their first job while the other 425 had first job tenures that were right censored at various points of time depending on when, after graduation in 1970, they had entered employment.[3] Figure 1 plots the max-

---

[2] A full description of the survey is contained in Williamson (1981) and further details of the variables used here are given below and in Dolton and Makepeace (1986).

[3] This is because our observations are all made in 1977 and people entered their first jobs at various dates between 1970 and 1977.

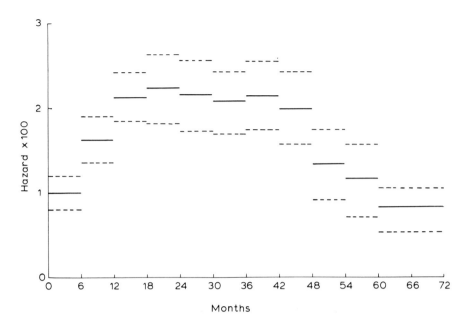

Fig. 1. Piecewise constant job tenure hazard (N = 1482).

imum-likelihood estimates of the pieces $\{\theta_j\}$; it constitutes a non-parametric description of the data. The dashed lines indicate the estimate of each piece plus or minus two estimated standard errors. Note that the estimated pieces are stochastically independent which explains the irregularity of the plot.

It is clear that the separation probability, or hazard function θ is non-monotone, increasing from near zero[4] to a peak at about 2 years and then falling. After 6 years the value of the function is about 0.008 implying a 0.8% probability of separation in any one month or about 10% per year. If people continued to leave at that rate the average employee would stay for a further (1/0.008) months or about 10 years, giving a completed tenure of 10 + 6 = 16 years. Thus after 6 years it is fair to say that the separation rate had become very low. It is, however, still positive implying, on the matching model, that learning about match quality was continuing after six years which seems a little implausible.

4.2

In the Jovanovic model separation occurs when the stochastic process m(t) crosses the absorbing barrier λ(t). The exact density function of this first

[4] This is confirmed when we fit models with more pieces and break up the 0–6 months piece into smaller components. These do indeed rise monotonically from near zero at t = 0.

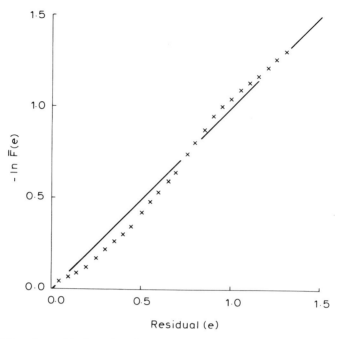

Fig. 2. Minus log residual survivor function inverse gaussian (2) hazard – job tenure data (N = 1482).

passage time can be deduced if we are willing to make some fairly strong functional-form assumptions – see Jovanovic (1984b). In particular if the distribution of match quality, μ, over matches is $N(0, \sigma_\mu^2)$ and if ε is an $N(0, \sigma_\varepsilon^2)$ Wiener process then the perceived match quality m(t) is a standard, N(0, 1), Wiener process on the transformed time scale p(t), where

$$p(t) = \sigma_\mu^2 \, rt/(1+rt)$$

where $r = \sigma_\mu^2/\sigma_\varepsilon^2$. If, in addition, the barrier is linear in the scale of p

$$\lambda(t) = -a + \alpha p(t); \quad a>0; \quad \alpha \geq 0$$

then the time to separation is the first passage time distribution of Brownian motion in the presence of a single absorbing barrier, in the scale of p. This is the Inverse Gaussian distribution with density function

$$f(p) = ap^{-3/2}\varphi\left[\frac{a-\alpha p}{\sqrt{p}}\right]$$

where φ is the standard Normal p.d.f. Of course f(t) is got by multiplying f(p) by $|dp/dt|$. This density function for t has no positive moments and is in fact defective, with a fraction of all entrants never leaving. Its hazard rises from zero to a unique maximum then falls to zero as $t\rightarrow\infty$, $p\rightarrow\sigma_\mu^2$. In spite of the fact

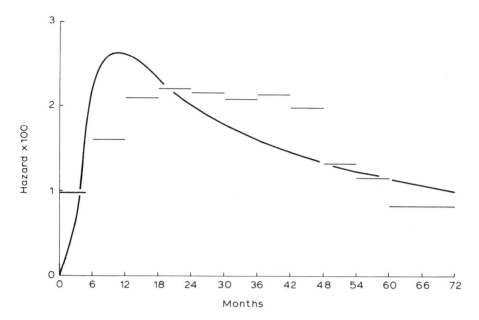

Fig. 3. Inverse gaussian and piecewise constant hazards.

that the p.d.f. involves 4 parameters, a, α, σ_μ^2 and σ_ε^2 only 3 parameters may be identified. We choose to normalise by setting $r\,\sigma_\mu^2 = 1$ so that

$$r = 1/\sigma_\mu^2$$

which is the reciprocal of the variance of the distribution of quality over matches. We now have

$$p(t) = t/(1 + rt).$$

In terms of the matching model we must have $r>0$ but if we set $r = 0$ as the limiting case we have that t itself is two-parameter Inverse Gaussian, i.e. the IG(2) model is nested within the three-parameter version. The results of fitting the IG(2) model are depicted in Figs. 2 and 3. Fig. 2 gives the Kaplan-Meier estimate of the minus log generalised residual survivor function which should plot approximately on a 45° line if the model is correct,[5] while Fig. 3 superimposes the estimated IG(2) hazard function on the piecewise constant hazard estimates of Fig. 1. Both plots indicate the fit is inadequate.

The first step in fitting the three-parameter model is to score test the null hypothesis $r = 0$. We find an (asymptotic) $N(0, 1)$ test statistic of -2.15 implying that a negative value of $r = 1/\sigma_\mu^2$ is consistent with the data. When we subsequently attempt to fit the three-parameter I.G. model the ML estimates

---

[5] See Lancaster and Chesher (1985).

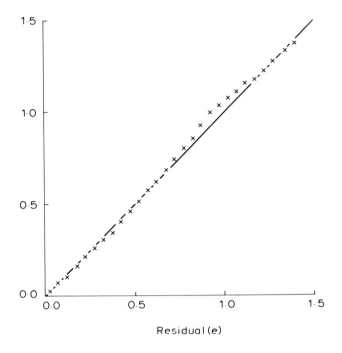

Fig. 4. Minus log residual survivor function plot log quadratic hazard – job tenure data (N = 1482).

of r approach (meaningless) negative values. Thus both three and two parameter Inverse Gaussian models are inconsistent with the data.

4.3

It is of interest to consider briefly a second parametric model, in which the hazard function is

$$\theta(t) = \exp\{\alpha_0 + \alpha_1 \ln(t+1) + \alpha_2 \ln^2(t+1)\},$$

the log quadratic model. If this model is to rise to a unique maximum we must have $\alpha_2 < 0$ is which case the integrated hazard behaves like $\int e^{-\alpha_2 \ln^2 s} \, ds$ for large t, which converges. Thus the model, like the three-parameter Inverse Gaussian, is defective when the hazard looks like Fig. 1. The ML estimates of the three parameters are

$$\alpha_0 = -7.72 \ (0.40)$$
$$\alpha_1 = 2.66 \ (0.27)$$
$$\alpha_2 = -0.45 \ (0.04)$$

Fig. 4 gives the Kaplan-Meier plot while Fig. 5 superimposes the estimated

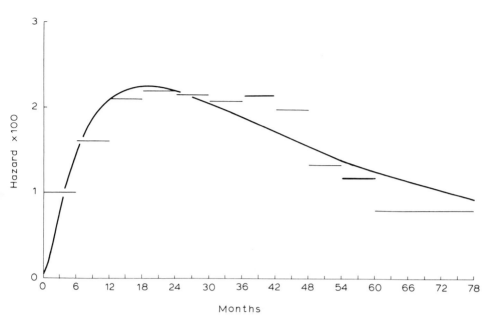

Fig. 5. Fitted log quadratic hazard superimposed on the piecewise constant estimator – job tenure data (N = 1482).

hazard function on the piecewise constant estimates. It appears that this model provides a tolerably accurate summary of the data.

5. Regression effects

In the simplest version of the matching model the separation probabilities do not depend upon the starting wage. In this section we report the effects of two regressors:

x_1 The logarithm of the annual starting salary in the first job deflated to 1970 base

x_2 The occupational social-status index of the first job according to the index of Goldthorpe and Hope (1974).

The wage is the real annual pre-tax salary. It does not, unfortunately, include the value of 'fringe benefits'. The occupational status variable is intended to capture a dimension of non-pecuniary remuneration.

In general the estimates of the effect of regressors on the hazard depend upon what is assumed about the functional form of the time-dependence of θ. Below we report two estimates of the coefficients β in models of the form $\theta = \exp\{x'\beta\}\theta_0$ for various assumptions about θ_0.

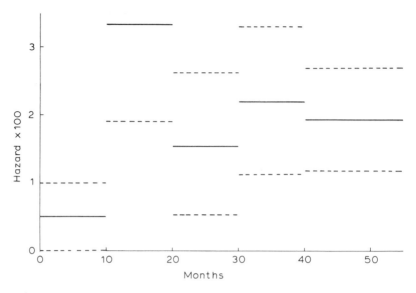

Fig. 6. Piecewise constant job tenure hazard civil, structural and mechanical engineering (N = 79).

Model for θ_o	Starting wage	Occupational status
Piecewise constant	−0.12 (.12)	−.008 (.005)
Log quadratic	−0.11 (.12)	−.007 (.005)

These results are consistent with the view that job-separation probabilities do not depend upon the starting wage; that is, they do not depend upon what is known about the productivity of the match at the moment it is formed.

6. Narrower occupational categories

It is arguable that it is more meaningful to classify people into more narrowly defined occupations than 'Private Industry and Commerce'. When we do classify people more narrowly by their occupation we find that the typical shape of Fig. 1 is retained. Fig. 6 for example gives the corresponding plot for entrants to 'Civil, Structural and Mechanical Engineering' with 79 observations of which 18 were right censored. The fitted log quadratic hazard is

$$\log \theta = -9.49 + 3.59 \ln(t+1) - 0.55 \ln^2(t+1) + 0.82 \ln X_1$$
$$(2.09)(1.36)(0.21)(0.46)$$

where t is time in months and X_1 is the starting wage. This hazard rises from

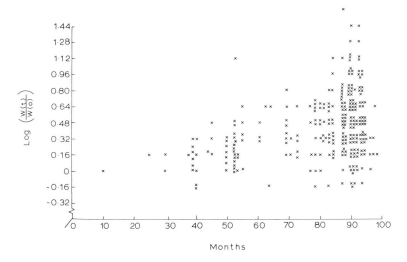

Fig. 7. Proportionate wage growth against tenure.

close to zero at the origin to a maximum at about 25 months, then falls. We examined four narrowly defined occupational groups for which there were sufficient observations and the same general shape was observed along with a non-significant coefficient on the starting wage. A formal test of homogeneity of the hazard functions for the four occupational categories

1. Chemists
2. Computer Programmers and Systems Analysts
3. Civil, Structural, and Municipal Engineers
4. Electronic Engineers

was performed by fitting separate models to each category and then fitting the same model to the pooled data. For the piecewise constant hazard models with five pieces plus one regressor – the log starting wage – the sum of the log likelihoods from the four separate fits was −1031.1 while the log likelihood from the pooled fit was −1039.4. This gives minus twice the log likelihood ratio of 16.6 on 18 degrees of freedom – (4 × 6 − 6). The null hypothesis of homogeneity could not be rejected. A similar result was obtained when the hazard was assumed to be log quadratic. Apparently there is no great harm in treating the observations as homogeneous between occupations.

7. Wages of survivors

Under the pure matching model the wage paid depends upon the initial measured characteristics, $g(x)$, which determine the initial wage, and upon the perceived match quality, say m. Thus

$$w(t) = g(x) + m(t)$$
$$w(0) + m(t).$$

Since those whose perceived match quality is low leave, survivors are composed increasingly of those whose m is large so m(t) is an increasing function of time (tenure). Thus

$$E[w(t) - w(0)] = f(t)$$

for some increasing function f(·).[6] This regression of wage growth on tenure can be investigated since we know the starting wage of all people and the wage being earned in 1977 – the survey date. At that data 377 people were still employed in their first job and reported their current wage and they had job tenures of various lengths depending upon how soon after graduation they had taken employment. Taking annual wages in 1970 £'s, measured logarithmically as w(t) and w(0), Fig. 7 gives the scatter diagram of w(t) − w(0) on tenure t. It is apparent that the longer people have been employed the larger on average is their wage (given their starting salary) and also the more variable is the amount of wage growth. If we fit a regression of wage growth on tenure, through the origin, the result is

$$w(t) = w(0) + .0068t.$$
$$(.0010)$$

Holding constant the starting wage an extra month's tenure adds, on average, an extra 0.68% to the current wage. In terms of the matching model this is a pure sample selection effect representing the tendency of survivors to be increasingly made up of those whose matches are productive. There is no significant non-linearity in the plot.

8. Conclusions

We have studied the probabilities of separation from their first job of British University graduates in the 1970's. These probabilities form a non-monotone function of time, rising from near zero to a peak at about two years then falling. This is true whether we consider entrants to jobs in all private industry and commerce taken together or whether we classify them into much narrower occupation categories. There is no evidence that probabilities of job separa-

---

[6] The exact form of the density function of w(t) given w(0) and t can be deduced if we are prepared to adopt the assumptions of section 4.2 – see Jovanovic (1984a) Indeed an exact *joint* p.d.f. of w(t) and t can be written down from which all four parameters of the matching model can in principle, be recovered. In view of the results of section 4.2 we did not attempt to fit such models in either joint (t, w(t)) or conditional (w(t)/t) form.

tion depend upon the starting wage. We have also studied the real wages of employees as a function of their job tenure. These wages increase with tenure from their initial values in an approximately linear manner and with increasing dispersion over time.

While the general shape of the hazard function is consistent with separation occurring as the outcome of a learning process as in the job-matching model, it seems clear that the eventual rate of decline of the separation probabilities is too slow for matching to be a fully satisfactory account of the separation process. Note that heterogeneity in the data would create separation probabilities that fall *faster* than those of homogeneous populations so the slow rate of decline we observe could not be attributable to this effect.

The growth of wages as a function of tenure seems also to be inconsistent with pure matching. It is true they do increase as the matching model predicts but the increase is roughly linear and the increasing dispersion is not consistent with a selection process in which the survivors become increasingly homogeneous with regard to their perceived match quality and hence with regard to their wages.

There are many ways forward. One which seems appealing would be to construct an empirically testable model which integrates both matching and optimal on the job search along the lines of Jovanovic (1984a). To do this really requires information about the destinations of job leavers, in particular their wages in the second job, as well as the wage they earned immediately prior to leaving.

Bibliography

Chesher AD and Lancaster T. 1985. Residuals, tests and plots, with a job-matching illustration. *Annales de l'INSEE,* 59/60, 47–70.
Dolton PJ and Makepeace GH. 1986. Sample selection and male-female earnings differentials in the graduate labour market. Forthcoming in *Oxford Economic Papers.*
Goldthorpe J. and Hope K. 1974. *The social grading of occupations.* Oxford: Clarendon.
Jovanovic B. 1979. Job matching and the theory of turnover. *Journal of Political Economy,* 87, 5:972–990.
Jovanovic B. 1984a. Matching, turnover and unemployment. *Journal of Political Economy,* 92, 1: 108–122.
Jovanovic B. 1984b. Wages and turnover: a parameterisation of the job-matching model, in: *Studies in labor market dynamics,* edited by G.R. Neumann and N.C. Westergaard-Nielsen. Berlin: Springer-Verlag, 158–167.
Williamson P. 1981. *Early careers of 1970 Graduates.* Department of Employment, Research Paper no. 26, London: Linneys Nansfield.

Estimating the intertemporal elasticity of substitution for consumption from household budget data

JOHN MUELLBAUER

1. Introduction

I take particular pleasure in celebrating through the theme of this paper Cramer's contributions to the study of household behaviour. Of these, I might specifically mention his 1957 paper which contains the first exposition of the neoclassical theory of the demand for durables, his 1962 book on the ownership of durables and his 1969 text. Particularly memorable for me from the latter's very nice exposition of Engel curves is his clear treatment of the identification problem in the Prais-Houthakker (1955) method of constructing household equivalence scales from Engel curves incorporating household composition effects. Though no great issues arise in the treatment of household composition effects in the Engel curves which I estimate below, I hope that he will enjoy the paradoxes of the paper that follows.

It is a remarkable and not generally known fact that by adopting simultaneously the general assumptions made by economists who estimate the consumption function and those who estimate demand systems, it is possible to estimate the elasticity of substitution between consumption in different periods from cross-section budget data. In other words, on these assumptions it is possible to draw inferences about savings behaviour merely by observing how households allocate their budgets to different commodities (of which saving is *not* one). Moreover, some British evidence examined below suggests that for Britain, the elasticity of intertemporal substitution is slightly above unity. Since apparently we do not observe a great deal of substitution in time, on aggregate data, see Hall (1985) and Deaton (1985), this seems at first sight a surprisingly high estimate. However, little observed substitution in aggregate might well be explained by aggregation phenomena, see Muellbauer (1986), rather than by individual preferences being in favour of a low degree of substitutability. In addition, habits or adjustment costs could help to explain the smoothness of the path of consumption.

Section 2 substantiates the theoretical claim. Section 3 presents empirical evidence while Section 4 draws conclusions.

2. A common set of assumptions for the life-cycle hypothesis and for demand systems

The following assumptions are commonly made under the life-cycle hypothesis for the consumption function. There is a single financial asset with an expected nominal rate of return equal to r_j in period j. Financial markets are perfect in that the household can lend or borrow at the same rate of return r_j. Labour-and-transfer income expected to equal y_j in period j is exogenous to the household. The following assumption is common to the life-cycle hypothesis and demand systems: consumption is defined as expenditure on nondurables or as a service flow from a durable good.

Given the assumptions made so far, the life-cycle budget constraint can be written in the present-value form

$$W_t = \sum_{j=t}^{T} \sum_{i=1}^{n} \hat{p}_{ij} q_{ij} \tag{1}$$

where $\hat{p}_{ij}$ is the discounted expected price for period j = $p_{ij} / \prod_{s=t}^{j} (1 + r_s)$ and W_t is the present value over the life-cycle of financial and human wealth, the latter being the discounted expected value of labour and transfer income.

In demand systems it is almost invariably assumed that preferences are separable between periods so that the within-period allocation decision can be made separately from the between-period allocation decision. Usually, it is assumed also that preferences over goods in each period are separable from leisure so that restrictions or decisions on the latter have no bearing on the allocation of a budget between different goods. In the life-cycle consumption hypothesis it is further assumed that between-period consumption decisions can be made independently of the amount of leisure in each period, ie. separability of the life-cycle utility function in goods from leisure.

Given these assumptions on preferences, the life-cycle utility function relevant for decisions on purchases of goods takes the form

$$u = V[v_t(q_{1t}, \ldots, q_{nt}), \ldots, v_T(q_{1T}, \ldots, q_{nT})] \tag{2}$$

where q_{ij} is consumption of good i planned for period j and T is the end of the household's time horizon.

Let us define expenditure in period j

$$x_j = \sum_{i=1}^{n} \hat{p}_{ij} q_{ij}. \tag{3}$$

Let

$$v_j = \psi_j(x_j, \hat{p}_j) \tag{4}$$

be the period j indirect-utility function which results from maximizing $v_j = v_j(q_i)$ subject to (3) w.r.t. the vector q_j, where $\hat{p}_j$ is the period j price vector. Then the between-period allocation problem can be written in the form

$$\max u = V[\psi_t(x_t, \hat{p}_t), \ldots, \psi_T(x_T, \hat{p}_T)] \tag{5}$$

$$\text{s.t.} \quad W_t = \sum_{j=t}^{T} x_j.$$

In the life-cycle hypothesis it is invariably assumed (see for example Modigliani and Brumberg (1955), Ando and Modigliani (1963)) that an increase in life-cycle wealth W_t is allocated in the same proportion to each period's planned consumption as existing wealth. One set of conditions under which this result is obtained is to assume
a) each period's price vector is proportional to the current period's
b) the utility function is homothetic in a certain sense.
This can be explained as follows. Let $\hat{p}_j = \theta_j \hat{p}_t$ where θ_j is a scalar which measures the general level of prices expected in period j relative to period t, i.e., $\theta_t = 1$. Then

$$v_j = \psi_j(x_j, \theta_j \hat{p}_t) = \psi_j(x_j/\theta_j, \hat{p}_t) \tag{6}$$

since the indirect-utility functions are homogeneous of degree zero in total expenditure and prices. Then (5) takes the form

$$\max u = V[\psi_t(x_t/\theta_t, \hat{p}_t), \ldots, \psi_T(x_T/\theta_T, \hat{p}_T)] \tag{7}$$

$$\text{s.t.} \quad W_t = \sum_{j=1}^{T} \theta_j(x_j/\theta_j). \tag{8}$$

If V[ ] in (7) is homothetic in the arguments x_j/θ_j, i.e., the real consumption indicator for period j, the problem of maximizing (7) subject to (8) is that of maximizing a homothetic utility function subject to a linear budget constraint with the θ's being treated as the prices. The solution takes the form for all j

$$x_j = k_j(\theta_t, \ldots, \theta_T; \hat{p}_t) W_t \tag{9}$$

where $k_j(\ )$ is linear homogeneous in θ's and zero-degree homogeneous in $\hat{p}_t$.

There are three difficulties with this derivation. Firstly, the assumption that households expect future relative prices to be the same as current ones is extreme. There are systematic reasons for relative prices to alter, for example, because of technological trends and changes in primary commodity prices which are to some extent predictable. It seems plausible that intelligent households would take these into account and not assume that present relative prices will always persist. Secondly, relative prices do change so that in period $t+1$, $k_{t+1}(\ )$ would depend not on $p_t(=\hat{p}_t)$ but on p_{t+1}. Such relative-price effects are almost invariably assumed absent in studies of the consumption function. Thirdly, and perhaps most fundamentally, when (7) is assumed to be homothetic, $\hat{p}_t$ is taken to be a 'nuisance' parameter vector which is treated as constant. But in general, when preferences are viewed from the perspective of $t+1$ or some other date, unless relative prices are *actually* unchanged, homotheticity of (7) given $\hat{p}_t$ is *not* the same as homotheticity of (7) replacing $\hat{p}_t$ by another price vector. It does not seem very satisfactory to assume that preferences alter in such a way that every period they are homothetic when viewed from the perspective of that period's relative prices.

Such difficulties are absent under a set of assumptions on preferences which have sometimes been made. These are that preferences are homothetic within periods as well as between periods. Then the indirect-utility functions take the form

$$v_j = \psi_j(x_j/b_j(\hat{p}_j)) \tag{10}$$

and

$$u = V(\psi_t(x_t/b_t), \ldots, \psi_T(x_T/b_T)) \tag{11}$$

is homothetic in x_j/b_j. Note that the latter condition is independent of the prices which go into each b_j. Treating x_j/b_j as real consumption and the b_j's as the prices, maximizing (11)

$$\text{s.t.} \quad W_t = \sum_{j=t}^{T} b_j(x_j/b_j)$$

gives

$$x_j = k_j(b_t, \ldots, b_T) W_t \tag{12}$$

which is the form typically assumed in studies of the life-cycle consumption function.

However, the assumption that within-period preferences are homothetic is quite unacceptable to anyone studying the allocation of total current expenditure to detailed commodities. Engel's law which states that the budget elasticity of food expenditure is less than unity is one of the few empirical regularities on which the evidence is in universal agreement – see Houthakker (1957).

Neither set of assumptions considered so far yield a life-cycle consumption function of the form (12) while at the same time permitting non-homothetic preferences within periods. However, there is a third set of assumptions which does yield both (12) and demand systems of reasonable generality. This makes use of a theorem by Gorman (1959), see also Blackorby, Primont and Russell (1978) for a statement and proof, and Deaton and Muellbauer (1980) for a discussion.

Let us transpose Gorman's 1959 'price-aggregation theorem' to the life-cycle context. In general the solution to

$$\max U(q_{1t}, \ldots, q_{nt}; \ldots; q_{1T}, \ldots, q_{nT})$$

$$\text{s.t.} \quad W_t = \sum_i \sum_j \hat{p}_{ij} q_{ij} \tag{13}$$

is $q_{ij} = G_{ij}(\hat{p}, W_t)$.

Gorman (1959) is interested in conditions under which the demand for expenditure $x_j = \sum_i \hat{p}_{ij} q_{ij}$ can be written in the form

$$x_j = G_j(b_t(\hat{p}_t), b_{t+1}(\hat{p}_{t+1}), \ldots, b_T(\hat{p}_T), W_t) \tag{14}$$

where $b_j(p_j)$ is an increasing linear homogeneous function for all j.

Gorman's price-aggregation theorem

(14) is a solution to (13) under the following three alternative conditions:
(i) within-period homotheticity and intertemporal weak separability, or
(ii) within-period preferences whose indirect-utility function can be written in the form

$$u_j = f_j(x_j/b_j(\hat{p}_j)) - h_j(\hat{p}_j) \tag{15}$$

together with intertemporal additive separability, or
(iii) preferences displaying a mixture of (i) and (ii).

Using Gorman's Theorem we can prove the following results:

Lemma

(12) is a solution to (13) for non-homothetic within-period preferences if the between-period utility function is additively separable and the within-period indirect utility function takes the form

$$u_j = (x_j^\alpha + a_j^\alpha)/(H_j^\alpha a_j^\alpha - a_j^\alpha) \tag{16}$$

for $\alpha \leq 1$, where $a_j(\hat{p}_j)$ is linear homogeneous and $H_j(\hat{p}_j)$ is zero-degree homogeneous.

(16) is the PIGL form of preferences which, see Muellbauer (1975, 1976), has powerful consequences for exact aggregation over households.

Proving the Lemma is straightforward:

Proof

(12) is a special case of (14) so that Gorman's price-aggregation theorem applies. Since we require non-homotheticity within periods, we can exclude cases (i) and (iii). From Gorman's theorem, we know that preferences have to be additively separable between periods and that the indirect-utility function within periods must be in the class (15). Thus the household's maximization

problem is

$$\max u = \sum_j \varphi_j(u_j) = \sum_j \varphi_j\{(x_j/b_j) - h_j\}$$

$$\text{s.t.} \quad W_t = \sum_{j=t}^{T} x_j.$$

This is equivalent to

$$\max u = \sum_j \varphi_j\{f_j(Z_j) - h_j\} \tag{17}$$

$$\text{s.t.} \quad W_t = \sum_{j=t}^{T} Z_j b_j \text{ where } Z_j = x_j/b_j.$$

For this to give demands of the form (12), we must have (17) homogeneous in Z_j and concave. This can only happen if $\varphi_j\{\ \}$ is linear, all j and if $f_j(\ )$ is homogeneous of degree $\alpha \leq 1$, all j. Thus $f_j(\ )$ is a power or log function. Without loss of generality we can write the cost function as

$$x_j^\alpha = b_j^\alpha(u_j + h_j) \tag{18}$$

for $\alpha \leq 1$.

However, this does not have the log case as the limit when $\alpha \to 0$ nor does it remain increasing in utility for $\alpha < 0$. Rewriting (18) in the PIGL form

$$x_j^\alpha = a_j^\alpha + u_j((H_j a_j)^\alpha - a_j^\alpha) \tag{19}$$

where $b_j^\alpha = a_j^\alpha(H_j^\alpha - 1)$ and $h_j = (H_j^\alpha - 1)^{-1}$,

gives

$$\ln x_j = \ln a_j + u_j \ln H_j \tag{20}$$

when $\alpha \to 0$ and remains well behaved for $\alpha < 0$. The cost function (19) is equivalent to the indirect-utility function (16).

To discuss the behavioural implications of (19), note that by Shephard's lemma, the cost function $x = c(u, p)$ has the property $\partial c(u, p)/\partial p_i = q_i$: so that $\partial \ln c(u, p)/\partial \ln p_i = w_i$, where w_i is the budget share of the ith good. Thus

$$x_j^\alpha w_{ji} = a_j^\alpha a_{ji} + u_j[(H_j a_j)^\alpha (H_{ji} + a_{ji}) - a_j^\alpha a_{ji}] \tag{21}$$

where $a_{ji} = \partial \ln a_j/\partial \ln \hat{p}_{ij}$, $H_{ji} = \partial \ln H_j/\partial \ln \hat{p}_{ij}$.

Using (16), (21) implies for $\alpha \neq 0$

$$w_{ji} = A_{ji} + B_{ji} x_j^{-\alpha} \tag{22}$$

where $A_{ji} = [(H_j a_j)^\alpha (H_{ji} + a_{ji}) - a^\alpha a_{ji}]/[(H_j a_j)^\alpha - a_j^\alpha]$

and $B_{ji} = a_j^\alpha(1 - A_{ji})$.

Analogously for (20)

$$w_{ji} = a_{ji} + \left(\frac{\ln x_j - \ln a_j}{\ln H_j}\right) H_{ji} \tag{23}$$

$$= A'_{ji} + B'_{ji} \ln x_j \tag{24}$$

for appropriate values of A'_{ji}, B'_{ji}.

Note that if $\alpha = 1$

$$\hat{p}_{ij} q_{ij} = A_{ji} x_j + B_{ji} \tag{25}$$

which represents the case of linear Engel curves. However, if $\alpha = 1$, the intertemporal-utility function is of the (linear) perfect-substitutes form, a most unrealistic assumption to make about intertemporal preferences! More generally, the elasticity of substitution between consumption in different periods is $1/(1 - \alpha)$ so that the parameter α governs both the shape of the Engel curves and the degree of substitution over time. Thus, the log case (24) which corresponds to $\alpha = o$, implies an intertemporal elasticity of substitution of unity. If $\alpha = -1$ so that the Engel curve take the quadratic form

$$\hat{p}_{ij} q_{ij} = A_{ji} x_j + B_{ji} x_j^2, \tag{26}$$

the elasticity of substitution is 1/2. Intuitively, it would be surprising if the elasticity of substitution were much above unity, suggesting α not much above zero. Given that the aggregate consumption function depends on expectations of the future which are not directly observable, it is difficult to identify the elasticity of substitution from such aggregate time series data. The difficulties are arguably less for budget data on which Engel curves can be estimated.

3. Estimating α from family expenditure survey data

The data used refer to the 5 years from 1968 to 1972 and to 11 commodity groups. Households in each demographic category are grouped by normal household income. Average expenditure on each of the commodity groups and average total expenditure for each normal income bracket give the budget shares and total expenditure used in estimating the Engel curves. The commodity groups are the following: food, fuel and light, housing, alcohol, tobacco, clothing and footwear, durables, miscellaneous goods, private transport, public transport and services. Housing includes the imputed rent of owner occupiers but durables expenditure refers to purchases rather than imputed rent. For the kind of grouped data under consideration, one would expect measured purchases of durables to be positively correlated with the unobserved rental equivalent but clearly purchases are the wrong concept.

This issue comes clearly out of the first neoclassical treatment of the demand for consumer durables due to Cramer (1957).[1]

Two responses are possible to this problem. One is to argue that the bias in the estimated curvature of the Engel curves is small: on the average the share of durables expenditure is less than 10%. The other is to omit durables and to interpret the theory developed in Section 2 to refer only to non-durable consumption expenditure. This would imply that the life-cycle utility function is separable in non-durables from durables as well as homothetic so that the proportionality hypothesis remains valid with k_j now depending on prices of durables as well as on those of nondurables.

One feature of Engel curves of the PIGL type (22) (or (24)) is that according to this preference hypothesis, the degree of commodity disaggregation is irrelevant for the estimation of α. One simple method of estimating α is to take a two-commodity breakdown, say food and nonfood, and estimate a share equation of the type (22) for food by non-linear least squares. This will actually be done below and the results compared with those derived from the full 10 commodity breakdown (excluding durables).

Furthermore, results will be presented for each of the 5 years separately. If there are important missing dynamic features from behaviour not represented by (22), for example the durability of goods or costs of adjustment, one might have thought that different years might yield somewhat different estimates of α.

For each year data are pooled across 4 demographic household types. The first household type consists of an adult couple without children in which the household head is below 65 years of age. The other types contain an adult

Table 1. Estimates of α from the food share equation.

	No. of obs.	Including durables			Excluding durables		
		α	2 ln L	χ^2	α	2 ln L	χ^2
1968	21	0.25	169.5	1.7	0.2	163.4	0.7
1969	20	0.3	170.2	4.7	0.2	175.0	2.1
1970	20	0	159.8	0	−0.2	162.7	0.3
1971	24	0.5	202.8	7.9	0.4	198.5	3.9
1972	28	0.85	207.0	14.2	0.85	204.6	13.1
1968–1970	61	0.25	498.3	5.2	0.15	499.4	1.4
1968–1972	113	0.40	900.6	19.8	0.3	894.4	12.0

[1] The further possibility of adjustment costs raises analogous problems for conventional budget data which excludes information on previous states in which the individual households comprising the sample found themselves.

couple and respectively, one, two and three or more children. The following demographic variables are introduced.

$D_1 = 1$ if the household contains one child, otherwise zero.
$D_2 = 1$ if the household contains two children, otherwise zero.
$D_3 = 1$ if the household contains three or more children, otherwise zero.
$D_4 =$ ratio of no. of children under 5 to the total no. of children if at least one child is present.

To avoid potential difficulty if α turns out to be close to zero, (22) is estimated in the form (omitting the j subscript)

$$w_i = A'_i + B'_i \left(\frac{x^{-\alpha} - 1}{-\alpha} \right) \tag{27}$$

since $\lim_{-\alpha \to 0} \left(\frac{x^{-\alpha} - 1}{-\alpha} \right) = \ln x$, and where both A'_i and B'_i vary demographically. The latter is achieved by writing

$$A'_i = a_{io} + \sum_{s=1}^{4} a_{is} D_s \tag{28}$$

and

$$B'_i = b_{io} + \sum_{s=1}^{4} b_{is} D_s. \tag{29}$$

Table 1 presents estimates of α for the 5 years from the foodshare equations including and excluding durables from total expenditure. These are obtained from OLS for a grid for α with 0.05 intervals. The estimates are very similar whether durables are excluded from total expenditure or not. However, they do show some variation from year to year, with a tendency for α to be higher in 1972. Surprisingly, apart from 1970, all the estimates are positive and this is supported by the results in Table 2 which bring to bear information from 10 commodity groups. Given that (27) holds for all commodity groups, the

Table 2. Estimates of α from a system of 10 commodity groups excluding durables.

	α	2 ln L	χ^2
1968	0.45	1437.8	6.3
1969	0.20	1438.3	4.1
1970	−0.30	1401.9	2.9
1971	0.35	1626.5	4.8
1972	0.30	1790.2	2.5
1968–1972	0.23	7683.5	9.4

estimates in Table 2 are more efficient. Incidentally, given the large number of parameters viz. 91, the estimates and likelihoods in Table 2 assume a diagonal covariance matrix.

One might ask whether there was anything special about 1970. This was a year when consumption and real income resumed the growth which had been interrupted from the last quarter of 1967 to the last quarter of 1969. Since $\alpha = 1$ for linear Engel curves, the estimates suggest the largest degree of curvature (ie. deviation from 1) in the Engel curves for 1970. It may be that an element of stock rebuilding in household expenditure accentuates the curvature of the Engel curves.

Be that as it may, the asymptotic chi-squared statistics in Tables 1 and 2 test the hypotheses $\alpha = o$ for each year. This can be accepted for 1968 and 1970 for both definitions of total expenditure and also for 1969 for total expenditure excluding durables. The lack of correction for degrees of freedom tends to bias these tests in favour of rejection. To get some idea of this, a pseudo F-statistic for 1969 when total expenditure includes durables computed as if the model were linear in α gives $F = 2.3$ with 1 and 9 degrees of freedom. The critical value at 5% is 5.1 so that the hypothesis is easily accepted while the asymptotic test rejects it, given chi-squared of 4.7 with a critical value of 3.8. On a pseudo F-test, the hypothesis $\alpha = o$ can be accepted for both definitions of total expenditure for 1968–1970 but rejected for 1968–1972.

The other hypothesis that can be tested from Tables 1 and 2 is equality of α for each year. The critical value of chi-squared is 9.5 with 4 degrees of freedom at 5%. The computed values of chi-squared are respectively 8.7 and 8.5 for Table 1 and 11.2 for Table 2. Given the asymptotic nature of the tests, this suggests that the hypothesis can be accepted. The point estimate of $\alpha = 0.23$ from Table 2 suggests an intertemporal elasticity of substitution for consumption expenditure excluding durables of 1.3. The asymptotic test rejects a value of unity. It is uncertain whether an appropriate degrees of freedom correction would give the same result. However, since the degrees of freedom are substantially higher than for the single equation results of Table 1, it seems likely that the elasticity of substitution does significantly exceed unity.

One of the chief difficulties of drawing conclusions from cross-section Engel curves concerns the problem of errors in the measurement of expenditure which come from the brevity of the sampling period. In the Family Expenditure Survey this is two weeks and over such a short period one cannot neglect the storability of goods even such as food and other reasons for low frequencies of purchase. Liviatan's (1961) suggestion of grouping on the basis of normal income as an instrumental variables technique for overcoming the measurement problem has been adopted in this study. However, there is an aggregation problem which arises when working with grouped data. Using the results of Muellbauer (1975, 1976), the appropriate total per capita expenditure variable in (27) for a particular normal income bracket is k x̄ rather than mean per capita expenditure x̄. k is an index of equality which is unity when there is zero inequality within the group and greater than unity in general. Unless the degree of equality happens to be the same in each group, the use of x̄ instead of k x̄ will cause consistency of the estimated parameters to be violated.

To illustrate, suppose the definition of the normal income brackets is designed to put about 20% of the total number of households into each of 5 brackets. With the mean roughly at the 60% level, one would expect k for the bottom quintile and the top two quintiles to be above that in the middle two quintiles. This would introduce a systematic bias in the estimated degree of curvature of the budget share equation. Intuition suggests the bias in α to be an upward one but without using the individual household data there is no reliable way of discovering what it is. However there is evidence in Cramer (1969), p. 151–3, which suggests that the bias is unlikely to be large. Cramer compares estimates of budget elasticities for linear, isoelastic and semi-log Engel curves estimated across 173 individual households and for the same data grouped into respectively 29 and 7 income classes. The estimates are very similar across the different aggregations of the data and, if anything, more similar for the non-linear Engel curves than for the linear ones! Recently, Kay, Keen and Morris (1982) have estimated linear Engel curves using normal income as an instrument for total expenditure on individual household data. The instrumental variables technique can also be extended and applied to the case of non-linear Engel curves and this may be worth doing in future work.

4. Conclusions

This paper has shown that the parameter α which governs the curvature of Engel curves also governs the intertemporal elasticity of substitution. This arises from the Gorman Price-Aggregation Theorem which is used to reconcile the proportionality hypothesis almost universally adopted by economists studying the life-cycle consumption hypothesis with the non-homotheticity of preferences within periods which Engel curves strongly support and which is almost universally adopted by economists studying demand systems. Evidence from U.K. Engel curves suggests a point estimate of α about 0.23 implying an intertemporal elasticity of substitution for individual households of about 1.3. This is not an unreasonable size of parameter though perhaps rather higher than those familiar with econometric results from aggregate time-series would have expected. But those results could be a consequence of aggregation phenomena.

If one takes the theory seriously, one can interpret it as providing an explanation for the empirical failure of linear Engel curves reported by many who have studied shapes of Engel curves, for example Prais and Houthakker (1955) and Leser (1963), and confirmed by the results reported above. Linear Engel curves imply perfect intertemporal substitution in this theory, an obviously unacceptable restriction on behaviour. The Working (1943) – Leser (1963) Engel curve, see (24) above, in which the budget share is a linear function of log-total expenditure and which underlies both the Nearly (Almost) Ideal Demand System and Jorgenson and Lau's (1975) Translog System, implies $\alpha = 0$ and hence an intertemporal elasticity of unity. Leser's empirical evidence, which favoured this specification relatively to all others he considered, is quite consistent with the small value of α estimated above. Its implication of a reasonable value of the intertemporal elasticity of substitution is rather satisfying.

On the other hand, perhaps preferences are more complex than these simple economists' characterizations. Perhaps the proportionality between consumption and life-cycle wealth suggested by time-series data is an aggregation phenomenon rather than a property of individual preferences. If so, while the Engel curves estimated above may be perfectly acceptable representations of cross-section behaviour, they would have no implications for savings behaviour. I remain agnostic on this issue but hope that this exploration of if ... then ... has maintained the reader's interest.

5. Acknowledgement

This research was supported by the E.S.R.C. under the Programme in Quantitative and Comparative Macroeconomics, grant number HR 6235. I am grateful to Lionel Mendis for programming the computations and to Terence Gorman for comments. Responsibility for errors is mine.

Bibliography

Ando A and Modigliani F. 1963. The 'life-cycle' hypothesis of saving: aggregate implications and tests. *American Economic Review,* 53: 55–84.

Blackorby C, Primont D and Russell RR. 1978. *Duality, separability and functional structure: theory and economic applications.* New York: North-Holland Publishing Co.

Cramer JS. 1957. A dynamic approach to the theory of consumer demand. *Review of Economic Studies,* 24: 73–86.

Cramer JS. 1962. *The ownership of major consumer durables.* Cambridge: Cambridge University Press.

Cramer JS. 1969. *Empirical econometrics.* Amsterdam: North-Holland Publishing Co.

Deaton A and Muellbauer J. 1980. *Economics and consumer behaviour.* New York: Cambridge University Press.

Deaton A and Muellbauer J. 1980. An almost ideal demand system. *American Economic Review*, 70: 312–326.

Deaton A. 1985. *Life-cycle models of consumption: is the evidence consistent with the theory?* Invited paper for the Fifth World Congress of the Econometric Society, Cambridge, Mass.

Gorman WM. 1959. Separable utility and aggregation. *Econometrica*, 27: 469–481.

Hall RE. 1985. Real interest and consumption. NBER working paper 1694, August 1985.

Houthakker HS. 1957. An international comparison of household expenditure patterns commemorating the centenary of Engel's law. *Econometrica*, 25: 531–551.

Jorgenson, DW and Lau LJ. 1975. The structure of consumer preferences. *The Annals of Economic and Social Measurement*, 4: 49–101.

Kaŷ JA, Keen MJ and Morris CN. 1982. Consumption, income and the interpretation of household expenditure data. Unpublished ms., Institute for Fiscal Studies.

Leser CEV. 1963. Forms of Engel functions. *Econometrica*, 31: 694–703.

Liviatan N. 1961. Errors in variables and Engel curve analysis. *Econometrica*, 29: 336–362.

Modigliani F and Brumberg R. 1955. Utility analysis and the consumption function: an interpretation of cross-section data, in: *Post Keynesian Economics*, edited by Kurihara KD. London: George Allen & Unwin.

Muellbauer J. 1975. Aggregation, income distribution and consumer demand. *Review of Economic Studies*, 62: 525–543.

Muellbauer J. 1976. Community preferences and the representative consumer. *Econometrica*, 44: 979–999.

Muellbauer J. 1986. Uncertainty, liquidity constraints and aggregation in the consumption function. Ms. Nuffield College, February 1986.

Prais SJ and Houthakker HS. 1955. *The analysis of family budgets*. Cambridge: Cambridge University Press.

Working H. 1943. Statistical laws of family expenditure. *Journal of the American Statistical Association*, 38, 43–56.

Associated with an income distribution and a demand system is a multidimensional expenditure distribution

HENRI THEIL

1. The n-dimensional expenditure distribution

The conventional approach to consumption theory considers only one consumer. His preferences are represented by a utility function, which is maximized subject to a budget constraint. Under appropriate conditions this leads to a system of demand equations, one for each good.

On the other hand, statistical data are usually available on a per capita basis. To formulate a demand system at that level, we consider the demand systems of *all* consumers and then we aggregate over these consumers.

But we can also look at all consumers jointly without aggregation. Then income (an independent variable in a demand system) for all consumers simultaneously becomes a distribution: the income distribution. The implication is that the expenditures of the consumers on the n goods are described by an n-dimensional expenditure distribution, one dimension being reserved for each good. Three simple examples, all based on a lognormal income distribution, are given in Section 2.

2. Three examples

The simplest demand model uses constant elasticities. For good i it takes the form

$$\log E_i = A_i + B_i \log E + u_i \qquad (1)$$

where E_i is a consumer's expenditure on good i, E his total expenditure (or income), B_i the i th income elasticity, A_i a constant (possibly depending on prices), and u_i a random disturbance. Assume that $(u_1, \ldots, u_n)$ follows a multinormal distribution. If the income distribution is lognormal, the distribution of log E is normal. It is then readily verified that the n-variate expenditure distribution (the distribution of $E_1, \ldots, E_n$ over the consumers) is multivariate lognormal. Of course, we face the problem that the sum of n lognormal variates is not itself lognormal, thus contradicting the assumption of a lognormal income distribution, but this simply reflects the fact that the constant-elasticity demand model violates the adding-up condition.

No such problem arises for the addilog model,

$$\frac{E_i}{E} = \frac{c_i(p_i/E)^{\gamma_i} e^{v_i}}{\sum_{k=1}^{n} c_k(p_k/E)^{\gamma_k} e^{v_k}} \tag{2}$$

where p_i is the price for good i, v_i is a random disturbance, and c_i and γ_i are constant. By dividing (2) by the same equation, but with i replaced by j, and by then taking logarithms, we obtain

$$\log \frac{E_i}{E_j} = \text{constant} - (\gamma_i - \gamma_j) \log E + v_i - v_j \tag{3}$$

where the constant depends on p_i and p_j (for simplicity, we assume that all consumers pay the same price for each good). We conclude from (3) that if $(v_1, \ldots, v_n)$ is multinormal, the ratio of the expenditures on any two goods has a lognormal distribution.

Finally, assume that Working's (1943) model applies:

$$\frac{E_i}{E} = \alpha_i + \beta_i \log M + \varepsilon_i. \tag{4}$$

If $(\varepsilon_1, \ldots, \varepsilon_n)$ is multinormal, so is the budget share vector $(E_1/E, \ldots, E_n/E)$. Since the normal distribution does not guarantee that such shares are nonnegative, we shall have to replace it by a truncated normal distribution.

3. A simulation

We apply Working's model, (4), to food and nonfood (n = 2) in nonstochastic form ($\varepsilon_i = 0$). For i = food, we specify $\alpha_i = 0.16$ and $\beta_i = -0.15$; these values are in approximate agreement with estimates obtained from cross-country data when the U.S. per capita income is defined to be 1. Four sets of 10,000 normal variates are generated with different means μ and variances σ^2 shown at the top of Table 1. By taking antilogs we obtain four lognormal income distributions with means of the form $\exp\{\mu + \frac{1}{2}\sigma^2\}$. The first two lines of the table contain this population mean and the corresponding arithmetic average of the antilogs of the 10,000 drawings. The agreement is obviously close. The lowest level of per capita income, 0.064, corresponds approximately to India's per capita income; the highest, 0.223, to that of Brazil. The last row of the table contains the per capita budget share of food ($\bar{w}_F$); it decreases as we move from left to right (i.e., as we move to a higher per capita income).

The remainder of the table concerns expenditure inequalities as summary measures of the bivariate expenditure distributions of the experiment (see the Appendix for details). J is inequality of income (or total expenditure). For a lognormal income distribution, $J = \frac{1}{2}\sigma^2$; the observed values are rather close to this population value. We can apply the J measure separately to food expenditure (J_F) and to nonfood expenditure (J_N). Not surprisingly, the inequality of food expenditure is much smaller than that of nonfood expenditure. By weighting J_F and J_N by $\bar{w}_F$ and $1 - \bar{w}_F$, respectively, we obtain $\bar{J}$, which is always larger than J except when consumption expenditure by goods and by purchasing individuals are independent (in which case $J = \bar{J}$). Here there is obviously no such independence, because the poor spend more on food relative to nonfood than the rich. The excess $\bar{J} - J$ equals $\bar{I}$, which is about 10 percent of J in all four cases. We can write $\bar{I}$ as a weighted average of I_F and I_N (with weights $\bar{w}_F$ and $1 - \bar{w}_F$), where I_F measures the inequality of the food budget shares across individuals (similarly I_N for the nonfood budget shares).

Details on the above matters follow in the Appendix, but the results clearly suggest that the concept of an n-dimensional expenditure distribution provides a significant extension of consumption theory. Further extensions are worthwhile, such as error terms in the allocation model, multiplication coefficients in such models that vary randomly from one consumer to the next, and other measures of the n-dimensional distribution in addition to inequality measures.

Table 1.

	$\mu = -3$ $\sigma^2 = 0.5$	$\mu = -3$ $\sigma^2 = 1$	$\mu = -2$ $\sigma^2 = 0.5$	$\mu = -2$ $\sigma^2 = 1$
Per capita income:				
true	0.064	0.082	0.174	0.223
observed	0.064	0.083	0.175	0.221
J: true	0.250	0.500	0.250	0.500
observed	0.250	0.503	0.250	0.477
J_F	0.129	0.231	0.094	0.153
J_N	0.438	0.821	0.387	0.701
$\bar{J}$	0.273	0.551	0.275	0.528
$\bar{I}$	0.023	0.048	0.025	0.052
I_F	0.020	0.057	0.040	0.115
I_N	0.027	0.040	0.015	0.023
$\bar{w}_F$	0.534	0.458	0.384	0.315

4. Appendix

In the simulations we disregarded all trials which yielded a food budget share which is negative or larger than 1. There were no such cases for $(\mu, \sigma^2) = (-2, 0.5)$ and only one (a food share >1) for $(-3, 0.5)$, but there were more such cases for $\sigma^2 = 1$.[1]

The inequality measures displayed in Table 1 are all from Theil (1979).[2] Let $P = 10{,}000$ be the population size, M_p the income of person p, and $M = M_1 + \ldots + M_P$ total income. The inequality measure used is

$$J = \log P - \sum_{p=1}^{P} \frac{M_p}{M} \log \frac{M_p}{M} \quad (\log = \text{natural log}) \tag{5}$$

Once the income of person p has been randomly drawn, we derive his food and nonfood expenditures (M_{Fp} and M_{Np}) using Working's model. We define M_F and M_N as the sums over $p = 1, \ldots, P$ of M_{Fp} and M_{Np}, respectively. Then $\bar{w}_F = M_F/M$, while the extension of (5) to J_F and J_N is straightforward.

We have $\bar{J} - J = \bar{I}$, where $\bar{I}$ is the expected mutual information across goods and persons. When the goods are food and nonfood, $\bar{I}$ can be written as the sum over $p = 1, \ldots, P$ of

$$\frac{M_{Fp}}{M} \log \frac{M_{Fp}/M}{(M_F/M)(M_p/M)} + \frac{M_{Np}}{M} \log \frac{M_{Np}/M}{(M_N/M)(M_p/M)}.$$

We have $\bar{I} = \bar{w}_F I_F + (1 - \bar{w}_F) I_N$, where

$$I_F = \sum_{p=1}^{P} \frac{M_p}{M} \left(\frac{w_{Fp}}{\bar{w}_F} \log \frac{w_{Fp}}{\bar{w}_F} \right) \tag{6}$$

while I_N is similarly defined. In (6), $w_{Fp} = M_{Fp}/M_p$ is the food budget share of person p, satisfying $\sum_p (M_{Fp}/M_p) w_{Fp} = \bar{w}_F$. It is easily seen that (6) measures the inequality of the food budget shares in the same way that (5) measures the inequality of income.

---

[1] For $(\mu, \sigma^2) = (-3, 1)$ we have 45 cases of a food budget share exceeding 1 (but no negative food shares). For $(-2, 1)$ we have 10 cases of a negative food share and two cases with a food share exceeding 1.

[2] In Theil (1979) I considered the measurement of income by components of income; the case of components of total expenditure is entirely analogous. Also, the 1979 paper uses two different measures of inequality; the measure which is not used in the present paper (indicated by a prime such as J' in the earlier paper) is not defined when an expenditure is zero.

5. Acknowledgement

Research supported in part by the McKethan-Matherly Eminent Scholar Chair, University of Florida.

Bibliography

Theil H. 1979. The measurement of inequality by components of income. *Economics Letters*, 2: 197–199.
Working H. 1943. Statistical laws of family expenditure. *Journal of the American Statistical Association*, 38: 43–56.

B. Money

The theory and measurement of cash payments: a case study of the Netherlands

EDGAR L. FEIGE

1. Introduction

One of the more intractable problems in the area of monetary economics is the measurement of cash payments. In recent years, interest in cash payments has been revived as a direct result of their alleged role in lubricating the 'underground' economy. Because cash payments rarely leave a 'paper' trail, they are an effective medium of exchange for those seeking to avoid the payment of income or consumption taxes. But the importance of cash payments goes well beyond the issue of the underground economy. Indeed, the appropriate measurement of cash payments, and more particularly, the turnover or velocity of cash, is essential for the coherent development of monetary theory, and for measuring the effects of monetary changes on all macroeconomic activity.

Since the stock of currency (C), is well defined and easily measured, the estimation of total cash payments $(C \cdot V_c)$ reduces to the problem of estimating the number of times the average unit of currency turns over (V_c) in any given period. The issue can be traced back to Jevons who in 1875 wrote:

'I have never met with any attempt to determine in any country the average rapidity of circulation, nor have I been able to think of any means whatever of approaching the investigation of the problem, except in the inverse way. If we knew the amount of exchanges effected, and the quantity of currency used, we might get by division the average number of times the currency is turned over; but ... the data are quite wanting.'[1]

The implications of the intellectual barrier established by the absence of a viable method for estimating V_c can not be underestimated. For example, the modern version of the quantity theory of money,[2] although intellectually

---

[1] Jevons, W.E. (1875) *Money and the Mechanism of Exchange* quoted in R. Selden (1956).
[2] Friedman, M. (1956) *The Quantity Theory of Money*. A restatement in M. Friedman (1956).

rooted in Fisher's pioneering elaboration of the Equation of Exchange,[3] departs from Fisher's emphasis on monetary flows, and reformulates the theory in terms of the *stock* of money. The commonplace emphasis on monetary stocks in both monetary and macroeconomic theory rather than on *payment flows,* is both arbitrary and pernicious. Arbitrary, because it assumes that the work money does is proportional to the money stock; pernicious, insofar as it destroys the independent behavioral role of payment velocity in monetary theory.

The essential work done by money as the medium of exchange is directly proportional to the volume of payments effected by money, not to the stock itself, except, of course, in the limiting case where the payment velocity of money is constant. It is the absence of independent estimates of payment velocity, (V_c), that necessitates the assumption, implicit in fifty years of monetary theory, that the services of money are proportional to its stock. Moreover, without independent estimates of V_c, theorists have been forced to specify the theory of money demand from the limiting perspective of income velocity rather than payment velocity.

The central role of V_c was clearly recognized by Fisher, who also proposed an ingenious method for measuring this elusive magnitude.[4] Fisher argued that:

'The importance of such accurate determinations (of the velocity of currency) can scarcely be overestimated. When we know statistically the velocity of circulation of money we shall be in a position to study inductively the 'quantity theory' of money, to discover the significance of that velocity in reference to crises, accumulations of wealth, density of population, rapid transit and communication, as well as many other conditions. In fact a new realm in monetary statistics will have been opened.' p. 618.

Yet, despite Fisher's claims for the theoretical and empirical importance of the payment velocity of currency, and his early efforts to devise a method for calculating V_c, little further attention was given to the problem for almost sixty years. The intervening years did, however, produce a plethora of studies of[5] 'income velocity', often in the guise of studies of the demand for money. But, what is commonly known as 'income velocity', namely, the ratio of observed income to the money stock, is neither a correct measure of the number of times money is spent to purchase currently produced goods and services, no is it a useful theoretical construct for predicting the consequences of changes in the

---

[3] Fisher, I. (1911).
[4] Fisher, I. (1909).
[5] See Selden (1956) for an excellent review.

money supply on economic activity. What is remarkable, is the tenacity with which economists continue to cling to the 'income'-velocity concept despite Keynes' early warning, that income velocity is 'a hybrid conception having no particular significance'[6] whose use, 'has led to nothing but confusion.'[7] Perhaps the same can be said for our commonplace definitions of 'the money supply', (M1; M2; ... M_i), concepts whose usefulness are limited by their total neglect of the weights required for meaningful monetary aggregation, namely, the payment velocities of the media of exchange.

It was not until 1970, that Laurent (1970),[8] apparently unaware of Fisher's earlier work, examined the issue of currency payments, and established the rudiments of a method for estimating currency tranfers. This work might have been sadly ignored, had it not been for the fortuitous circumstance that it was found handy as a means for implementing the 'transactions' method[9] for estimating the size and growth of the 'underground' economy.

The purpose of this paper is to elaborate Laurent's conceptual framework in order to gain a greater understanding of both the theory and measurement of currency payments and more particularly, the payment velocity of currency. Section 1 reviews Fisher's early efforts to develop a procedure for estimating cash payments and currency velocity. This method, reincarnated by Cramer[10] has been employed by the Netherlands Central Bank in order to obtain estimates of currency payments in the Netherlands as part of a study that attempted to apply Feige's (1979; 1980) 'transaction' method for estimating the underground economy in the Netherlands. The procedures and results of the Netherlands Central Bank are evaluated and are later compared with the alternative set of estimates developed in Section 4. Section 2 reviews the essential features of Laurent's (1970) approach to the estimation of currency payments, and Section 3 elaborates this framework to produce a theoretical model of denomination-specific currency payments. The model incorporates institutional variables such as the currency-quality standard that is a decision variable of the monetary authority. It also incorporates engineering factors that pertain to the physical characteristics of the currency itself. Both of these theoretical innovations are relied upon when the model is empirically implemented in Section 4. The empirical section includes estimates of denomination specific cash payments in the Netherlands from 1950 to 1984. Empirical estimates are also obtained for the total number of lifetime transactions that a unit of currency can on average effect, as well as estimates of the average lifetime of each denomination and its payment velocity.

---

[6] Keynes, J.M. (1930), Vol. II, p. 24.
[7] Keynes, J.M. (1936), p. 299.
[8] Laurent, R. (1970).
[9] Feige, E. (1979).
[10] Cramer, J.S. (1981a).

The final section of the paper summarizes the results of the study and suggests further applications of the methodology to some fundamental issues of monetary and macroeconomics.

2. The Fisher cash loop method

At the beginning of the century, Fisher (1909) devised an ingenious approximation for estimating the velocity of cash payments. Fisher constructed a model of cash flows that described the various types of currency exchanges that take place between the time individuals initially withdraw cash from the banking system and subsequently redeposit it. Sectoring his hypothetical economy into firms, individual depositors and non-depositors, Fisher examined all possible cash exchanges between the three sectors, and then derived an expression for the value of cash payments effected in the acquisition of goods and services.[11]

Fisher assumed that individuals exclusively acquire cash by withdrawing it from banks.[12] Given information on the value of total withdrawals (ω), and an estimate of the number of payments that occurred between cash withdrawals and subsequent redeposits [the 'cash loop' (λ)], the value of cash payments (P) is simply:

$$P = C \cdot V_c = \omega \cdot \lambda \tag{2.1}$$

where:
P represents the value of cash payments.
C represents the stock of currency in circulation.
V_c represents the average velocity or turnover of currency.
ω represents cash withdrawals from the banking system.
λ represents the 'cash loop'.

Fisher's model specification was conditioned by the institutional structure of the payments mechanism as it functioned during the early years of the 20th century. The approximate formula he derived for the 'total circulation of cash in exchange for goods' equaled the total amount of cash deposited in banks plus total wages paid. Given this estimate of cash payments, and information on the stock of currency outstanding, Fisher estimated that the average velocity of currency was approximately 18 turnovers per year, implying an average holding period between cash exchanges of about three weeks, and a 'cash loop' (λ) approximately equal to 1.6.

---

[11] In the more general model, presented in Section 3, total cash payments include cash payments made for the acquisition of goods and services as well as the acquisition of real and financial assets.
[12] A model of currency payments that includes real and financial assets other than 'money' would also permit the acquisition of cash from sales of real and financial assets. Such 'sales' would of course include cash withdrawals from time and savings deposits.

Seventy years elapsed before Fisher's loop length estimate was utilized by Cramer (1981a)[13] in his efforts to measure the volume of cash payments in the Netherlands. Cramer obtained rough estimates of the volume of cash withdrawals in the Netherlands and applied Fisher's estimate of the 'cash loop' ($\lambda \approx 2$), to obtain estimates of cash payments in the Netherlands. This same procedure was subsequently adopted by the Netherlands Central Bank (1984) in its efforts to estimate the value of cash payments.[14] Boeschoten and Fase, (1984) obtained estimates of cash withdrawals from banks and giro institutions for several years and multiplied these estimates by the assumed 'cash loop' ($\lambda = 2$) to obtain currency payments. In 1980, this procedure yielded an estimate of average currency velocity of 15.25 implying that the average currency note is held for 24 days before being spent.

3. The simple average lifetime method

Laurent (1970) suggested an alternative framework for estimating the volume of cash payments (P) and hence, the velocity of currency (V_c). If it is assumed that notes can effect a total of G payments during their lifetime, namely, between the time a note is initially issued and finally withdrawn from circulation, and one can obtain an estimate of the average lifetime of notes at time t L(t), it follows that:

$$V_c(t) = G/L(t). \tag{3.1}$$

and

$$P(t) = V_c(t) \cdot C(t) = G/L(t) \cdot C(t). \tag{3.2}$$

Laurent was able to obtain empirical estimates of the average lifetime of currency notes from data on issues and redemptions of currency and could therefore identify changes in cash payments over time in terms of the unknown parameter (G).[15] In order to obtain estimates of (G), Laurent employed Fisher's equation of exchange:

$$C(t) \cdot \frac{G}{L(t)} + D(t) \cdot V_d(t) = [PT](t) \tag{3.3}$$

and assumed that:

$$\frac{[PT](t)}{[py](t)} = k_0 \tag{3.4}$$

[13] Cramer, J.S. (1981a).
[14] Boeschoten, W.C. and M.M.G. Fase (1984), pp. 20–21.
[15] In principle, G could be empirically estimated from well-designed engineering study that tracked note exchange in a controlled environment that simulated real-world conditions.

where:
D = Checkable deposits.
V_d = Velocity or turnover of checkable deposits.
PT = Value of total transactions.
py = Value of current income or product (GNP).
k_0 = Constant.

Assumption (3.4) enabled Laurent to substitute nominal income as a proxy for total transactions in the equation of exchange (3.3). He then selected that value of G which maximized the correlation between the two sides of equation (3.3). Laurent found that an estimate of G ≈ 129 produced the best fit for his modified equation of exchange.

Assumption (3.4) has been implicitly and explicitly utilized in monetary economics for more than half a century, and was rarely questioned prior to its use as a key assumption in the 'transactions' method for estimating the size and growth of unrecorded and unreported income.[16] Indeed, the assumption plays an important role in most studies of the demand for money that are specified in terms of income rather than transactions. However, the past decade has witnessed an unprecedented degree of financial innovation that may have significantly raised the ratio of transactions to income. Due to the paucity of data on gross financial flows for most nations, it is currently difficult to determine the extent to which the ratio has risen as a result of the understatement of income[17] (the unrecorded 'income hypothesis') or as the result of a vigorous increase in gross financial transactions. Regardless of how that issue is finally resolved, one is left with the uncomfortable conclusion that (3.4) is a tenuous assumption for inference in monetary economics. Unfortunately, the abandonment of this assumption calls into question much of the empirical literature on the demand for money that relies on (3.4) to permit the substitution of income for transactions in the money demand function. It will be recalled that the remarkable 'monetarist' counter-revolution rested on the cornerstone of a stable demand function for money. Yet in the United States, predictions from conventional money-demand functions became highly unreliable during the mid 1970's, precisely at the time when the rising ratio of (PT) to (py) was producing very high estimates of unrecorded and unreported income. It is possible, as we shall later suggest, that Laurent's use of assumption (3.4) was reasonable during a large part of the period he studied, but is no longer tenable.

[16] Efforts to estimate the value of total transactions [PT] have been undertaken by Cramer (1981a, 1981b) for the Netherlands and the United Kingdom, and by Feige (1980; 1981; 1985a) for the U.S.; the U.K. and Sweden.
[17] The Bureau of Economic Analysis has recently released a major revision of the U.S. National Income and Product Accounts that incorporates an upward revision of National Income of $147.5 billion for 1984 to take account of misreported income on tax returns.

A recent major survey of currency usage in the U.S. sheds some light on this issue. Undertaken by the Board of Governors of the Federal Reserve System,[18] the survey of household currency usage estimated that the average turnover of currency for product expenditures in 1984 was between 50 and 55 turnovers per year. Taking the lower figure, in conjunction with estimates of the average lifetimes of U.S. currency notes, Feige, (1986) estimated that $G \approx 155$ in 1984.[19]

Since this recent survey was specifically designed to obtain an estimate of currency velocity, its findings can be used to obtain estimates of the total number of lifetime transactions (G), that notes of various denominations can sustain before being judged unfit for further circulation. Given estimates of (G) for the U.S., and an estimated relationship between (G) and certain physical characteristics of paper currency as measured by folding test machines that are used in both the U.S. and the Netherlands, it is possible to infer a value of (G) appropriate for the currency of the Netherlands. These calculations are fully described in Section 5.

4. A denomination-specific model of currency velocity and cash payments

In order to gain the maximum benefit from the point estimate of the average velocity of currency obtained by the recent Federal Reserve survey, it is necessary to specify a denomination-specific model of currency payments over time. The average velocity of currency ($V_c(t)$) for any time period t, is defined as:

$$V_c(t) = \frac{P(t)}{C(t)} = \frac{1}{C(t)} \cdot \sum_{i=1}^{d} V_i(t) \cdot C_i(t) =$$

$$\sum_{i=1}^{d} V_i(t) \cdot w_i(t), \qquad i = 1 \ldots d. \tag{4.1}$$

Similarly,

$$V_i(t) = \frac{P_i(t)}{C_i(t)} \tag{4.1}^1$$

where:
P(t) is the total volume of payments effected by all denominations of currency during period t.

---

[18] Spindt et al (1985); Avery et al (1985).
[19] The earlier estimate of currency velocity employed by Feige, (1980) (based on the Laurent method modified by information pertaining to changes in the physical characteristics of notes) ranged between 55 and 61 turnovers per year. The Federal Reserve Board survey results thus lend credence to the Laurent procedure.

$V_c(t)$ is the average velocity of currency during period t.
$C(t)$ is the average stock of currency of all denominations held during period t.
$w_i(t)$ is the share of the ith currency denomination. ($C_i(t)/C(t)$).

The institutional setting for the theory of cash payments is characterized by a monetary authority that issues notes (I) to the banking system. Firms and individuals withdraw currency from the banking system and use the currency notes to effect payments for assets, goods and services. Currency notes are eventually redeposited with the banking system, which returns the notes to the monetary authority. The monetary authority, being responsible for maintaining the fitness standard of the outstanding currency supply, sorts the notes received from circulation, and determines which notes are fit for recirculation and which notes are unfit and thus require redemption (R). The redeemed notes are those, that upon inspection, are deemed to be sufficiently worn to warrant destruction. For given fitness standards and physical characteristics of currency notes at time t, there exists a total number of payments $(G_i(t))$[20] that notes of denomination i can make during their lifetime before being redeemed. The number of cumulative payments $(N^*_i(t))$ performed by notes of denomination i by period t is:

$$N^*_i(t) = G_i(t) \cdot \left\{ \int_0^t R_i^n(s) \cdot ds + \int_0^t \{I_i^n(s) - R_i^n(s)\} \cdot \gamma_i(s) \cdot ds \right\} \quad (4.2)$$

where:
$N^*_i(t)$ is the number of cumulative payments performed by notes of denomination i by t.
$G_i(t)$ is the total number of payments that notes of denomination i can make in their lifetime before being destroyed, given the fitness standards and physical characteristics of paper currency in existence at time t.
$I_i^n(s)$ is the number of new notes of denomination i issued during period s.
$R_i^n(s)$ is the number of notes of denomination i redeemed during period s.
$\gamma_i(s)$ is the average fraction of lifetime payments performed by notes of denomination i as of date s.

The cumulative value of payments effected by notes of denomination i up to period t is:

$$P^*_i(t) = G_i(t) \cdot \left\{ \int_0^t R_i(s) \cdot ds + \int_0^t \{I_i(s) - R_i(s)\} \cdot \gamma_i(s) \cdot ds \right\} \quad (4.3)$$

where:
$R_i(s) = R_i^n(t) \cdot D_i$; where D_i is the nominal value of denomination i (e.g. Fl. 5; 10; 25; 50; 100; 1000); and similarly,
$I_i(s) = I_i^n(t) \cdot D_i$.

[20] $G_i(t)$ can be viewed as the mean of the distribution of lifetimes.

The total cumulative sum of payments effected by all denominations by period t is then:

$$\sum_{i=1}^{d} P^*_i(t) = \sum_{i=1}^{d} G_i(t) \cdot \left\{ \int_0^t R_i(s) \cdot ds + \int_0^t \{I_i(s) - R_i(s)\} \cdot \gamma_i(s) \cdot ds \right\} \quad (4.4)$$

Combining equations (4.1) and (4.3) it is possible to express the velocity of any denomination i as:

$$V_i(t) = \frac{\frac{d}{dt}\left\{ G_i(t) \cdot \left\{ \int_0^t R_i(s) \cdot ds + \int_0^t \{I_i(s) - R_i(s)\} \cdot \gamma_i(s) \cdot ds \right\} \right\}}{C_i(t)} \quad (4.5)$$

If we assume that the average note outstanding has performed one half of its total lifetime payments, differentiation with respect to t yields:

$$V_i(t) = \frac{.5 \cdot G_i(t) \cdot [R_i(t) + I_i(t)] + \frac{\dot{G}_i(t)}{G_i(t)} \cdot P^*_i(t)}{C_i(t)} \quad (4.6)$$

where:

$$\dot{G}_i(t) = \frac{d}{dt} G_i(t).$$

The actual number of lifetime payments ($G_i(t)$) that notes of denomination i can effect, depends upon *both* the physical characteristics (Q^*_p) of the notes and the quality fitness standard (Q^*_s) established by the monetary authority. If (G^*) denotes the number of lifetime payments that notes with physical characteristics (Q^*_p) can sustain under the quality fitness standard (Q^*_s), then the effective number of lifetime payments ($G_i(t)$) that notes of denomination i can perform is:

$$G_i(t) = \delta(t) \cdot G^*(t) \quad (4.7a)$$

where,

$$E\{\delta_i(t)\} = 1 \quad (4.7b)$$

and

$$\delta_i(t) \begin{cases} = 1 \text{ if } Q_s = Q^*_s \\ < 1 \text{ if } Q_s > Q^*_s \\ > 1 \text{ if } Q_s < Q^*_s \end{cases} \quad (4.7c)$$

The monetary authority attempts to maintain the quality of currency in circulation by maintaining a quality fitness standard Q^*_s which is expected to prevail over time. Temporary departures from the fitness standard will change the effective number of lifetime transactions that, on average, units of currency can effect. For example, when the monetary authority wishes to introduce a new series of currency into circulation, it might temporarily raise fitness standards and retire old series notes before they have reached the normal level of deterioration. In this case, the actual fitness standard $Q_s > Q^*_s$ and $\delta < 1$, implying that old-series notes that have been prematurely redeemed will effect a smaller actual number of total lifetime transactions (G). Alternatively, if for reasons of economy, the monetary authority temporarily permits the quality of the outstanding stock of currency to deteriorate, $Q_s < Q^*_s$ and $\delta > 1$, and the effective number of lifetime transactions will increase, as notes, normally regarded as unfit for circulation, are recirculated into the payments system.

Incorporating the above specifications into (4.1), the average velocity of currency can now be written as:

$$V_c(t) = \sum_{i=1}^{d} w_i(t) \cdot \frac{\delta_i(t) \cdot .5 G^*_i(t)[R_i(t) + I_i(t)] + \left\{ \frac{\dot{G}_i(t)}{G(t)} + \frac{\dot{\delta}_i(t)}{\delta(t)} \right\} P^*_i(t)}{C_i(t)} \quad (4.8)$$

If we assume that:
a) $G^*_i(t) = G^*(t)$ for all i; namely, that all denominations have identical physical characteristics,
and,
b) $\dot{G}^*_i(t) = \dot{\delta}_i(t) = 0$; namely, that during the period t there are no changes in physical characteristics nor in fitness standards, then:

$$V_c(t) = \frac{1}{C(t)} \cdot .5 G^*(t) \cdot \sum_{i=1}^{d} \delta_i(t)[R_i(t) + I_i(t)]. \quad (4.9)$$

If it is also the case that the monetary authority maintains uniform fitness standards for all currency denominations, then;
c) $\delta_i(t) = \delta(t)$ for all i,
and:

$$G^*(t) = \frac{V_c(t)}{\frac{1}{C(t)} \cdot \delta(t) \cdot .5 \sum_{i=1}^{d} [R_i(t) + I_i(t)]} \quad (4.10)$$

Equations (4.9) and (4.10) are instructive insofar as they specify the precise set of variables required to obtain estimates of the average velocity of currency (V_c) and the total number of lifetime payments of currency notes (G^*). Data on the stock of currency ($C_i(t)$) are readily available as is information on the value of new issues ($I_i(t)$) and redemptions ($R_i(t)$) of currency. Quantitative information on fitness standards (δ) could in principle be supplied by the monetary authority since it uses highly sophisticated automated sorting machines whose fitness-standard calibration is both set and recorded by the monetary authority. Thus, estimates of currency velocity require some specific knowledge of G^*, the total number of payments that the average note can sustain during its lifetime. Conversely, an estimate of G^* requires an independent estimate of V_c.

Fortunately, the aforementioned Federal Reserve survey provides an independent estimate of V_c for the United States in 1984. If we assume that the quality of U.S. currency during that year was approximately at its normal level (i.e., $δ ≈ 1$), equation (4.10) can be used to obtain a direct estimate of G^* for the United States, since the necessary information on $C(t)$; $R_i(t)$ and $I_i(t)$ is readily available.[21]

In order to obtain a provisional estimate of V_c for the Netherlands, it is necessary to take account of any differences in the currency quality standards that are maintained in the two countries and any differences in the physical characteristics of the paper which makes up the currency issue of the two countries.

An exact measure of the relative currency quality standard maintained by the monetary authorities in the Netherlands and the United States could in principle be obtained by an engineering study that examined the differential percentage of notes sorted as unfit from a prearranged test deck of currency that was sorted by the soil-detection equipment used in each country.[22] In the absence of exact information, it was assumed that the average currency quality standard maintained in the Netherlands was 25% higher than that which obtained in the United States.

In order to compare the physical characteristics of the paper used to produce currency in the two nations, the study relies on test results from Schopper machines that both nations employ to conduct fold-endurance tests on their currency issue. The aim is first, to determine the functional relationship between estimates of G^*, and quantitative test scores (X) obtained by fold-endurance test procedures in the United States, and then to infer the appropri-

---

[21] Feige, E. (1986).
[22] The sorting machines used in the United States and the Netherlands are made by different manufacturers. On the basis of interviews with technical experts in currency – sorting technology, it was determined that differences in the calibration settings on currency – sorting equipment produce a currency quality standard in the Netherlands that is approximately 20–25% higher than that maintained in the United States.

ate G* for the Netherlands from fold test scores on Dutch currency.[23] Thus;

$$G^* = f(X) \tag{4.11}$$

and $f(X)$ is estimated to be the square root function displayed in Fig. 1.[24] Estimates of G* for the Netherlands were then obtained from information on (X) for Dutch currency notes. G* for the Netherlands is conservatively estimated to be 80. Taking account of the finding that the currency quality standard of the Netherlands is 25% higher than that of the United States implies that $G^* \approx 60$ for the Netherlands.[25]

Given a point estimate of G*, the final estimation of the time path of the velocity of currency requires an econometric specification of equation (4.8). We shall assume that the velocity of currency of the ith denomination depends upon a non-linear time trend (t); the opportunity cost of holding a unit of the ith denomination ($OC_i(t)$) and the convenience ($CO_i(t)$) of the ith denomination as a store of value. The use of a non-linear time trend reflects Fisher's insight that payment habits are rooted in institutional considerations and thus change gradually over time. The opportunity-cost variable is included to reflect the fact that as the opportunity cost of holding currency increases, individuals will attempt to economize on outstanding currency balances and meet transactions needs by increasing the number of turnovers of the smaller

---

[23] (X) is the fold-test score obtained on the Schopper machine under given test conditions. Test conditions are determined by temperature (T = 23° C) and humidity (H = 50%). Under these test conditions, U.S. currency notes currently score (X = 4000) on cross direction fold endurance tests. In the Netherlands, scores of (X = 1500) where obtained under test conditions of (T = 23° C) and (H = 65%) and (X>900) under test conditions identical to those used in the U.S. Since the higher humidity test conditions employed in some of the tests in the Netherlands impart greater flexibility to the paper, and thus would result in a somewhat higher (X) score, it was estimated that comparable (X) scores for the Netherlands under identical test conditions would yield $X \approx 1200$. I am indebted to L.A. Wolfe and J. Mercer of the Department of the Treasury, Bureau of Engraving and Printing for providing test scores and technical details on the production and testing of currency notes. Comparable data for the Netherlands are from the Unpublished Reports of the Statistics Committee of the Banknote Printers Association.

[24] β_1 is estimated by OLS from the three available data points on G* and X from U.S. data. The end points are: (0,0) and the current measure of $G^* \approx 155$ and X = 4000 described above. The third set of coordinates was obtained from the (1962) study by E.B. Randall Jr. and J. Mandel that concluded that the 1957 introduction of dry intaglio notes increased the lifetime of notes by 30% when compared with the pre-1957 wet intaglio notes. Schopper test scores (X) for the pre-1957 notes where obtained from the Bureau of Engraving and Printing, and a time series for G* was constructed by assuming that the new notes with longer lifetimes were introduced into circulation during the period 1957–1964.

[25] This estimate of $G^* \approx 60$ is considerably larger than that employed by the Netherlands Central Bank. (Boeschoten and Fase (1984). The Bank obtained its estimate by arbitrarily assuming that the 'cash loop' was two. Multiplying this cash loop by an estimate of withdrawals of cash from bank and giro institutions yields an estimate of cash payments and hence an estimate of $V_c \cdot G^* \approx 34$, in turn, is derived from equation (4.10).

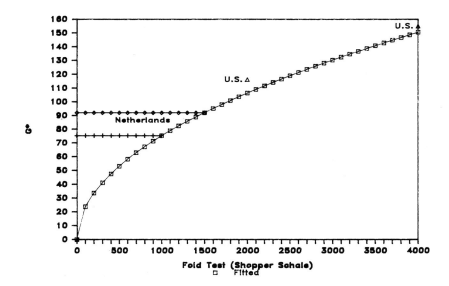

Fig. 4.1. Relationship between fold tests and G*.

average-currency balances. The opportunity cost is represented by:

$$OC_i(t) = 100 \cdot \left\{ \frac{PI_t}{PI_{t-1}} - 1 \right\} \cdot \frac{D_i}{PI_t} \qquad (4.12)$$

where,
PI = Price index.
Di = Nominal value of denomination i.

The convenience of the ith denomination as a store of value is assumed to vary directly with its nominal value (D_i) and may be influenced by the availability of other convenient stores of value ($A_i(t)$). In the Netherlands, it is widely believed that some unreported taxable income is hoarded in the form of interest-yielding bearer certificates of deposits issued by banks and giro institutions. In 1976, government officials publicly suggested that banks and giro institution might have to register bearer certificate sales, a policy that would increase the attractiveness of large denomination currency notes for hoarding. A dummy variable (A) is included in the regression for 1000 guilder notes to capture the short-term increase in payment velocity that might be expected as individuals substitute bonds for cash. The longer-run effect would be the opposite, namely to make 1000 guilder notes more attractive for hoards with lower turnover. Finally the regressions include a dummy variable (S) to capture the unusual movements in redemptions and new issues during periods where new-series notes are introduced or old series are prematurely withdrawn.

The foregoing considerations are expressed by letting:

$$Vc_i(t) = \exp\{D_i + \alpha_{i1}T(t) + \alpha_{i2}T^2(t) + + \alpha_{i3}S(t) + \\ + \alpha_{i4}A_i(t) + \alpha_{i5}OC_i(t)\}. \qquad (4.13)$$

Imposing the constraint that:[26]

$$\frac{\dot{G}^*(t)}{G^*(t)} = 0$$

and combining (4.9) with (4.13) yields:

$$\log\{T_i(t)\} = \log\{\Pi_i(t) - \frac{\dot{\delta}(t)}{\delta(t)} \cdot P^*_i(t)\} - \log\{\delta(t)\} \qquad (4.14)$$

where:

$$T_i(t) = G^*(t)[\{R_i(t) - I_i(t)\}\gamma_i(t)]/C_i(t)$$

$\Pi_i(t) =$ the rhs of (4.13).

Taking a first-order Taylor's expansion about $\Pi_i(t)$;

$$\log\{T_i(t)\} \approx \log\{\Pi_i(t)\} - \mu_i(t) \qquad (4.15)$$

where:

$$\mu_i(t) = \left\{ \frac{\frac{\dot{\delta}^*(t)}{\delta(t)} \cdot P^*_i(t)}{\Pi_i(t) \cdot C_i(t)} + \log\{\delta(t)\} \right\}$$

Given the assumptions (4.7b) and (4.7c):

$$E\,\mu_i(t) = 0,$$

and therefore (4.15) can be estimated by ordinary least squares.

The estimation results are reported in Tables 4.1 – 4.4. The most significant effect in the regressions is the constant term, confirming Fisher's notion that payment habits are deeply entrenched and are denomination-specific. The 1000-guilder equation reveals the expected short-term positive effect on velocity of 1000-guilder notes from the announcement that bearer bonds might require registration. The opportunity-cost variable is not significant in any of the equations, a result, explicable in terms of the relatively modest inflation experience of the Netherlands.[27]

---

[26] This constraint will only be violated in those years where changes in the physical characteristics of the currency issue are introduced. When such changes occur, the constraint is readily maintained by eliminating those specific years from the final estimation procedure.

[27] Manski and Goldin (1987) found that in the high-inflation environment of Israel, the opportunity cost variables significantly contributed to the explanation of the demand for currency.

Table 4.1. 1960–1981 dependent variable: C_{1000}.

Variable	Coefficient	Std. error	t-stat.	2-Tail sig.
D_{1000}	3.0221113	0.4458889	6.7777231	0.000
T	0.0732943	0.0535845	1.3678268	0.190
T^2	− 0.0027040	0.0013165	− 2.0539010	0.057
S	− 0.2331478	0.0884705	− 2.6353167	0.018
A_{1000}	0.2688564	0.1017147	2.6432408	0.018
OC_{1000}	− 0.0306460	0.1129781	− 0.2712562	0.790
R-squared	0.832516		Mean of dependent var	3.119706
Adjusted R-squared	0.780177		S.D. of dependent var	0.314522
S.E. of regression	0.147465		Sum of squared resid	0.347933
Durbin-Watson stat	2.064380		F-statistic	15.90625

Table 4.2. 1960–1981 dependent variable: C_{100}.

Variable	Coefficient	Std. error	t-stat.	2-Tail sig.
D_{100}	2.4802826	0.2468559	10.047490	0.000
T	0.0117898	0.0338228	0.3485743	0.732
T^2	0.0003144	0.0008339	0.3770760	0.711
S	− 0.4891320	0.0560489	− 8.7268861	0.000
OC_{100}	0.0693581	0.0711516	0.9747931	0.343
R-squared	0.870966		Mean of dependent var	3.039160
Adjusted R-squared	0.840605		S.D. of dependent var	0.234650
S.E. of regression	0.093682		Sum of squared resid	0.149198
Durbin-Watson stat	1.315872		F-statistic	28.68712

Table 4.3. 1960–1981 dependent variable: C_{25}.

Variable	Coefficient	Std. error	t-stat.	2-Tail sig.
C	2.7350857	0.2811110	9.7295574	0.000
T	0.0344256	0.0346537	0.9934169	0.334
T^2	0.0003141	0.0008543	0.3677150	0.718
S	− 0.2719165	0.0574308	− 4.7346795	0.000
OC_{25}	0.0577513	0.0728097	0.7931818	0.439
R-squared	0.924168		Mean of dependent var	3.681889
Adjusted R-squared	0.906325		S.D. of dependent var	0.313631
S.E. of regression	0.095991		Sum of squared resid	0.156642
Durbin-Watson stat	2.046867		F-statistic	51.79504

Table 4.4. 1960–1981 dependent variable: C_{10}.

Variable	Coefficient	Std. error	T-stat.	2-Tail sig.
D_{10}	3.4804807	0.2069485	16.818099	0.000
T	−0.0059402	0.0226697	−0.2620320	0.796
T^2	0.0009004	0.0005590	1.6106490	0.126
S	0.0879227	0.0375682	2.3403481	0.032
OC_{10}	0.0798636	0.0476787	1.6750374	0.112

R-squared	0.925874	Mean of dependent var	3.820583
Adjusted R-squared	0.908432	S.D. of dependent var	0.207537
S.E. of regression	0.062801	Sum of squared resid	0.067048
Durbin-Watson stat	1.204886	F-statistic	53.08443

Given the forecast values of V_c, it is now possible to utilize the foregoing framework to gain further insights into the role of currency payments in the Netherlands economy.

5. Empirical results

5.1 Denomination-specific results

On the basis of the estimated equations presented in Tables 4.1–4.4 it is now

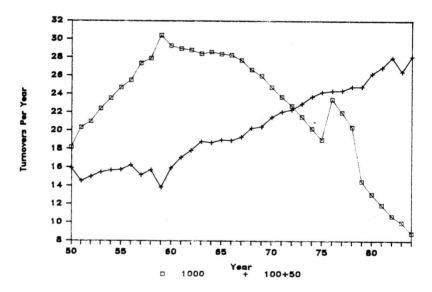

Fig. 5.1. Estimated currency velocity, large denominations.

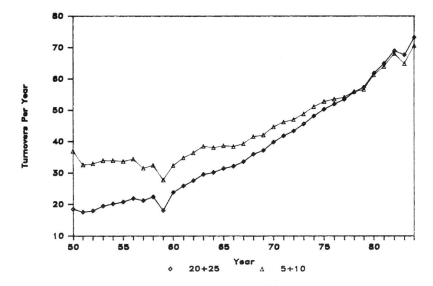

Fig. 5.2. Estimated currency velocity, small denominations.

possible to obtain the forecast values of the velocity of currency for each denomination over time. Figures 5.1 and 5.2 display estimates of currency velocity for large and small denominations respectively.

In general, one would expect to find that the velocity of currency varies inversely with the size of denomination. This expectation is confirmed for all denominations with the exception that the velocity of 1000 guilder notes appears to have exceeded that of 100 guilder notes during the period 1950–1972. By 1984, the average yearly turnover of 1000 guilder notes was 9 turnovers per year compared with 28.5 for 100 guilder notes and more than 70 turnovers for the smaller denomination notes. Over time, the velocity of currency for all notes other than 1000 guilders appeared to increase. The time path of the velocity of 1000-guilder notes suggests that during the decade of the 1950's, these large-denomination notes were widely used as a means of payment, however, during the past twenty five years, the 1000 guilder note was increasingly hoarded.

The average lifetime of notes in circulation, is expected to vary directly with denomination. As displayed in Figures 5.3–5.6 by 1984, 1000-guilders notes had an average lifetime of almost seven years in circulation, whereas, 100 guilder notes circulated 2.1 years and smaller denomination notes remained in circulation for approximately 10 months before being redeemed.

Figure 5.7 displays the estimated average holding period for each denomination. By 1984, small-denomination notes were held for approximately five days

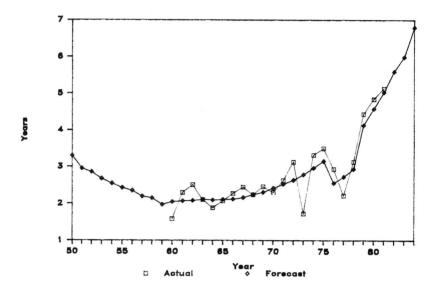

Fig. 5.3. Average lifetime of currency, 1000 guilder denomination.

before being spent, whereas, 100 guilder-notes were held approximately 13 days, and 1000-guilder notes were held approximately 42 days before being spent.

Given the differences in the inter-temporal and denomination-specific holding periods and velocities, it is apparent that the volume of payments per-

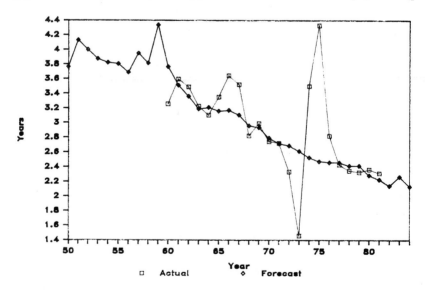

Fig. 5.4. Average lifetime of currency, 50 + 100 guilder denominations.

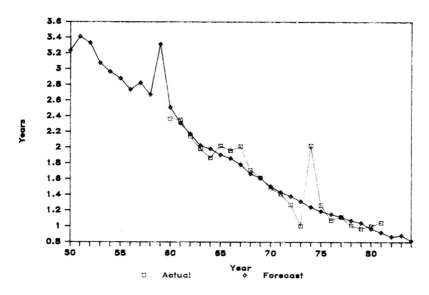

Fig. 5.5. Average lifetime of currency, 20 + 25 guilder denominations.

formed by different denominations of currency over time was affected not only by the amount of each denomination in circulation, but more importantly, by the intensity with which each denomination was used as a means of payment. Given the monetarist emphasis on money 'supply' in juxtaposition to the Fisherian emphasis on the work money does, it is instructive to examine each

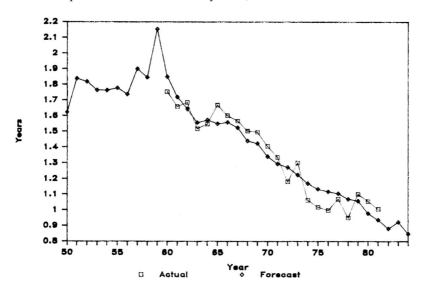

Fig. 5.6. Average lifetime of currency, 10 + 5 guilder denominations.

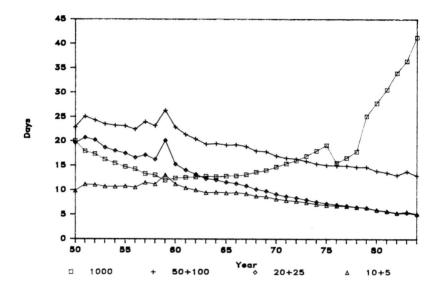

Fig. 5.7. Estimated average holding period, by denomination.

denomination's share of the total currency stock outstanding in comparison to each denomination's share of currency payments effected.

Figures 5.8–5.11 display the share of each denomination as a percent of the value of the currency stock as well as the share of each denomination as a percent of total cash payments.

As revealed by Figures 5.8–5.11 the share of each denomination as a percent of the currency stock is a poor indicator of the payments effected by each denomination. Thus, whereas 1000-guilder notes now make up almost 38% of the total value of the Netherlands currency stock, they account for less than 14% of total currency payments. Conversely, the smaller denomination notes, account for approximately 10% of the nation's currency supply but account for more than 30% of all currency payments.

5.2 Aggregate estimates of currency velocity and cash payments

Given the aforementioned denomination-specific estimates of currency velocity, it is now possible to obtain estimates of the average velocity of currency and the total volume of cash payments in the Netherlands. These estimates can in turn be compared with estimates of currency velocity and aggregate cash payments obtained by the Netherlands Central Bank by means of the simple Fisher cash-loop method.

The final estimates of the denomination-weighted average velocity of cur-

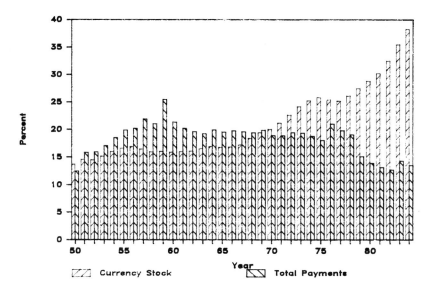

Fig. 5.8. Share of currency stock and payments, 1000 guilder denomination.

rency are displayed in Figure 5.12 as well as a recalculation of the Netherlands Bank's estimates (labeled B-F(1) and B-F(2)).[28]

The new estimates imply that the average unit of currency turns over approximately 25 times per year as compared with the estimate of approximately 15 turnovers per year calculated by Cramer (1981a) and by the Netherlands Bank, who obtained their estimates by assuming a cash loop length of two. According to the new estimates, the average unit of currency is held for approximately two weeks before being spent, compared with an average holding period of 3.5 weeks implied by the Netherlands Bank's estimates.

Figure 5.13 displays a comparison between the estimated average lifetime of currency calculated in the present study, and that employed by the Netherlands Bank's study. The figure reveals a relatively close correspondence between the two calculations except for those periods during which a new series of notes was issued, or an old series notes was withdrawn.[29] Over the period studied, the average currency note remains in circulation for approximately 2.8 years before being deemed unfit for further circulation.

Figure 5.14 displays the final estimates of G^* and $G = G^*\delta$. Recall, that the effective total number of payments that the average bill will sustain through-

---

[28] Series (B-F(1)) are based on $G = 34.3$ and (B-F(2)) are based on $G = 37$. Boeschoten and Fase, (1984), p 45.

[29] It will be recalled that the present analysis adjusts for these aberrations with the dummy variable (S), whereas the Bank's estimates employ the actual observed values of redemptions and issues to calculate average lifetime.

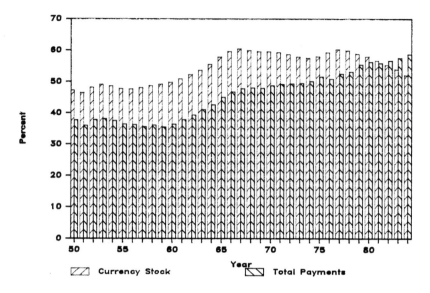

Fig. 5.9. Share of currency stock and payments, 50 and 100 guilder denominations.

out its lifetime (G(t)) depends upon *both* its physical characteristics and departures from the particular quality-fitness standard maintained by the monetary authority. Whereas the physical characteristics of currency remained constant over the period of study, fluctuations in the quality-fitness standard induce changes in the effective number of lifetime payments (G).

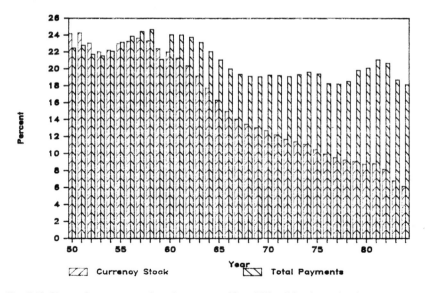

Fig. 5.10. Share of currency stock and payments, 20 and 25 guilder denominations.

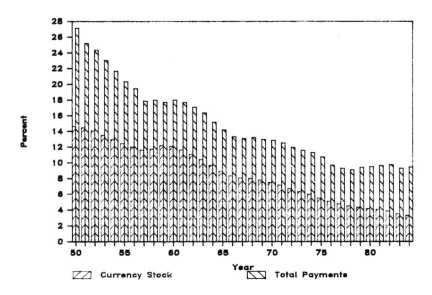

Fig. 5.11. Share of currency stock and payments, 10 and 5 guilder denominations.

The major difference between the findings of the present study and those obtained by Cramer (1981a) and the Netherlands Bank (1984) result from the treatment of the 'cash loop'. In the present study, the value of the 'cash loop' is derived from estimates of total cash payments, whereas in the earlier studies, estimates of cash payments are derived from an *assumed* value for the cash

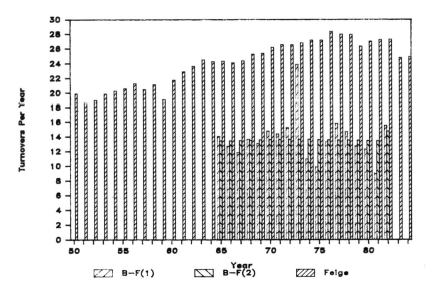

Fig. 5.12. Estimated average currency velocity, weighted by denomination share.

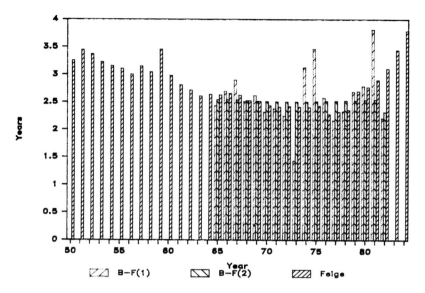

Fig. 5.13. Estimated average lifetimes, weighted by denomination share

loop. Figure 5.15 displays the difference between the estimated 'cash loop' from the present study, and, the assumed 'cash loop' in the earlier studies.

It appears from Figure 5.15 that the cash loop in the Netherlands is approximately equal to four, suggesting that currency is used to make four payments before being returned to the banks. The size of the cash loop may be explained

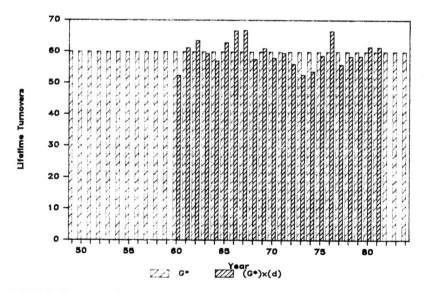

Fig. 5.14. Estimated physical and effective G, total lifetime turnovers.

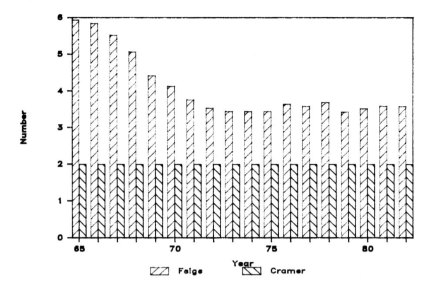

Fig. 5.15. Estimated and assumed cash loop.

by several factors. First, whenever currency is returned to the banking system, an audit trail is produced, since banks in the Netherlands make specific paper entries of all cash deposits and cash withdrawals. Thus, any firm or individual engaged in cash transactions that are not fully reported to the tax authorities has a strong incentive to recirculate the cash directly, rather than deposit it in the banking system. Furthermore, the Netherlands fiscal authority attempts to collect VAT taxes on final sales of goods and services. Individuals who effect transactions in cash, can frequently obtain merchandise and services without being charged the VAT tax. Sellers of merchandise and services can attract customers by agreeing to this practice, but in turn must 'skim' these cash payments from reported final sales. This in turn provides an incentive to hire workers who are willing to receive cash wages. Such workers can thus avoid income taxes, while firms can reduce their cost figures appropriately to match their reduced level of reported sales. Firms thus have a dual incentive to deal in cash. They can avoid employment taxes by paying 'off the books' workers in cash, and they can increase their competitive advantage relative to compliant firms by reducing final prices by the amount of the avoided VAT taxes. In some cases, firms have been known to collect VAT taxes on sales effected with currency, underreport their final sales and pocket the VAT tax collected. With VAT rates in the neighborhood of 18%, such incentives appear substantial.

One might also note that the cash loop is likely to be higher in countries with relatively low rates of theft, since one of the incentives to return currency to banks is security. The *Social and Cultural Report* (1980) indicates that convic-

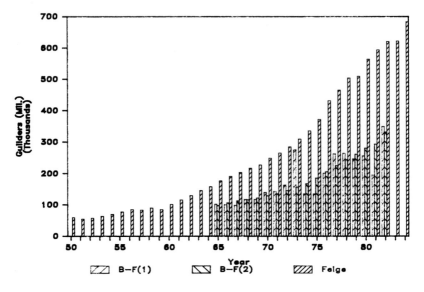

Fig. 5.16. Estimated currency payments

tions for offenses relating to property are less than 200 per 100,000 of the population aged 12–79 and, 'More than half of all recorded cases of simple theft relate to the theft of a vehicle' (p. 166). Thus, it appears that the security incentive to return currency to banks is unlikely to be compelling in the Netherlands.

Figures 5.16–5.19 provide information on the estimated volume of total cash

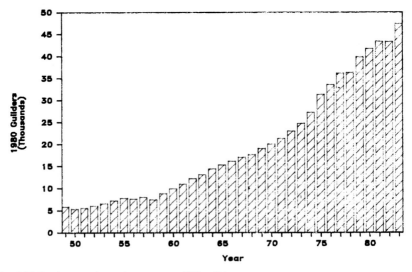

Fig. 5.17. Real per capita cash payments, 1980 guilders.

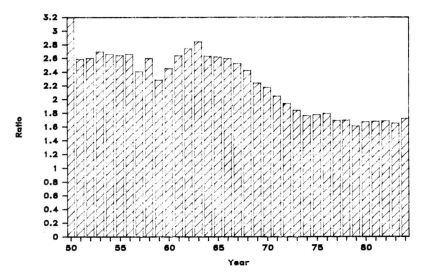

Fig. 5.18. Cash payments per guilder of GDP.

payments in the Netherlands. Figures 5.16 and 5.17 respectively reveal that cash payments have increased steadily throughout the period of estimation in both nominal and real per capita terms. Figure 5.16 reveals that the nominal estimates of cash payments of the present study are more than 1.7 times higher than those estimated by the Netherlands Bank (labeled (B-F(1) and (B-F(2)). As displayed in Figure 5.17 per capita cash payments exceeded 40,000 guilders in 1980. Of this total, approximately Fl. 5000 of payments were made with 1000-guilder notes, Fl. 25,000 of payments were effected with 50 and 100 guilder notes and the remaining Fl. 10,000 were effected with small denomination notes. As displayed in Figure 5.18 cash payments amounted to more than 1.7 times recorded GDP.[30]

In order to gauge the plausibility of these estimates, it is necessary to recall that cash payments represent all transactions that are effected by the use of currency. Such payments would include the sum of: final purchases of goods and services; factor-payment transactions; intermediate goods and services purchased; transfer payments; purchases of existing assets, both real and financial; purchases of newly created financial assets; purchases of foreign goods and services and purchases of foreign assets. On the basis of estimates of total payments made with checks and giro transfers[31] it is possible to estimate

---

[30] The foregoing calculations assume that the total stock of guilders in circulation are held by residents of the Netherlands. To the extent that some fraction of the Netherlands currency supply is held by residents of other countries, the estimates of total cash payments in the Netherlands will be overstated.

[31] Boeschoten and Fase (1984) pp. 62–63.

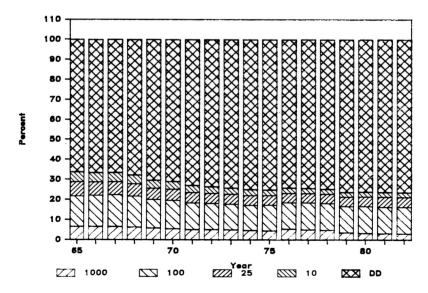

Fig. 5.19. Shares of payments.

the share of total payments (MV) made up by currency payments.

Figure 5.19 reveals that currency payment's share of estimated total payments (MV) declined from 33.7% in 1965 to 23.6% in 1982. The corresponding currency share of the money supply (currency + demand deposits and bank and giro institutions) declined from 45.7% in 1965 to 34% in 1982 (Figure

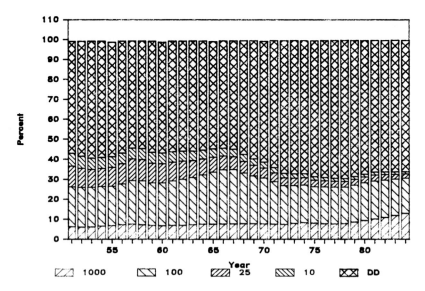

Fig. 5.20. Shares of the money supply.

5.20). In short, the role of currency in the monetary system has declined over time as measured by both stock and payment shares. Nevertheless, we are far from the 'cashless' society that many economists predicted. Moreover, gauging the role of currency and its components by stock measures, gives as inflated picture of the true role of currency in the payments mechanism.

5.3 The relationship between total payments and total transactions

In light of Fisher's claim that knowledge of the velocity of currency and cash payments could open 'a new realm in monetary statistics', it seems useful to explore some of the possible applications of the foregoing estimates of cash payments.

From the perspective of macroeconomic accounting, the measurement of cash payments makes possible the independent measurement of total payments (MV), suggesting the possibility that the equation of exchange (MV = PT) can be employed to provide a higher-order accounting identity than the conventional income-expenditure identity (Y = C + I + G), that has served the profession for the past fifty years.[32] Total transactions (PT), can be viewed as an aggregate that encompasses the key statistical entries in all current macroeconomic accounting schemes, in particular, the national-income and product accounts (NIPA); the input-output accounts (IO); the balance-of-payments accounts (BOP), and the flow-of-funds accounts (FOF).[33] Thus, the separate measurement of total monetary payments (MV) provides an independent empirical check on an appropriately aggregated sum of the entries in all of our present accounting systems. The long sought after, but highly elusive, 'integrated system of economic accounts', finds a natural conceptual framework in the flow equality constraint between payments and transactions. From this perspective, a positive difference between MV and PT, has a natural interpretation, namely, the sum of 'unrecorded' transactions and the 'statistical' discrepancy.

The Netherlands Central Bank has compiled an aggregative series for total transactions (PT),[34] that permits a preliminary examination of the question, 'Does MV = PT?'.

As displayed in Figure 5.21, the difference between estimated total payments and estimated total transactions[35] is both large and growing over time.

[32] See Feige, E. (1985a).

[33] In principle, total transactions must also include transfer payments and 'gross' financial transactions. Since past attempts to integrate current accounting systems have focused on balance sheet constraints rather than flow constraints, many financial flows are currently only available on a 'net' basis.

[34] Boeschoten and Fase (1984).

[35] Cash withdrawals from demand deposits are added to recorded transactions since such withdrawals give rise to demand deposit debits that are included in MV.

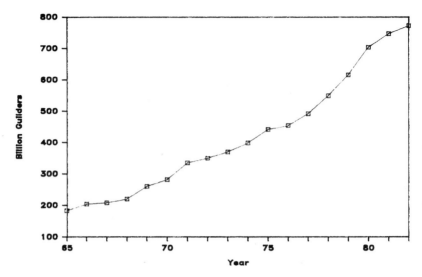

Fig. 5.21. MV–PT.

This sizable discrepancy between estimated payments and estimated transactions suggests that there remains a substantial volume of gross financial transactions that have not been properly accounted for in our current measures of PT. It would therefore be premature to conclude that the observed discrepancy between recorded payments and transactions represents an indicator of the size and growth of 'underground' transactions and hence unrecorded incomes. The problem, however, does not lie with the conceptual framework for estimating 'underground' transactions as suggested by the Netherlands Bank study, but rather with the dearth of empirical information concerning gross financial transactions. The equation-of-exchange constraint on our national accounting systems calls attention to this serious gap in our empirical knowledge. Only after this empirical deficiency is remedied, can we proceed to utilize Fisher's identity to obtain more accurate estimates of underground economic activities.

6. Summary and conclusions

Keynes' biographical essay on Robert Malthus ends with the lament:

> 'If only Malthus, instead of Ricardo, had been the parent stem from which nineteenth-century economics proceeded, what a much wiser and richer place the world would be today'.[36]

[36] Keynes, J.M. (1956), p. 36.

On this 50th anniversary of the publication of Keynes' *General Theory,* one is tempted to ask whether the twentieth-century world of economics would have been a wiser and richer place had its parent stem been Fisher rather than Keynes. Fisher's equation of exchange is clearly a more general conceptual framework than the income-expenditure approach that became the mainstay of Keynesian analysis. Yet, the equation of exchange did not benefit from the conceptual and empirical elaboration enjoyed by the income-expenditure approach. The economics profession's failure to pursue Fisher's payment-transaction approach was at least partly the result of the lacuna created by the inability to estimate the velocity of currency and hence cash payments. This paper has attempted to fill that gap by specifying a theory and a means of estimating currency velocity.

Given the means for measuring currency velocity and cash payments, it is possible to derive total payments estimates and to compare these with independent measures of total transactions. The equation of exchange can then be utilized as a higher-order constraint on the sum of all monetary transactions in the economic system. Once payments and transactions can be independently estimated, the equation of exchange can serve as the conceptual framework for integrating the monetary payments mechanism with an integrated system of currently available transactions accounts.

The ability to measure denomination-specific currency velocities also frees monetary economics from its long-standing tradition of measuring monetary stocks as the nominal sum of the existing media of exchange. Aggregate quantities of goods and services have always been measured by index numbers that take account of the relative value of different goods and services as measured by their market prices. Given estimates of the velocities of the different media of exchange, it now becomes possible to appropriately measure aggregate monetary stocks as index numbers that take account of the relative work that different media of exchange perform in the payments system. The appropriate weights for monetary aggregation are the velocities of the different media used for exchange. Given velocity-weighted measures of the 'money supply', and conversely, money-supply-weighted measures of payment velocities, it becomes possible to formally test the underlying economic relationships between monetary aggregates and final, intermediate and financial transactions and prices. In short, Fisher's equation of exchange can provide the conceptual and empirical basis for the development of a general theory of economic transactions.

7. Acknowledgement

The initial research for this paper was undertaken while the author was a

Fellow at the Netherlands Institute for Advanced Studies and subsequently as Cleveringa Professor at the University of Leiden. Financial support from the A.P. Sloan Foundation and the Graduate School of the University of Wisconsin is gratefully acknowledged as is the assistance of Mark Kennett.

Bibliography

Avery R et al. 1985. Cash and transaction account usage by American households. *Board of Governors of the Federal Reserve System.*

Boeschoten WC and Fase MMG. 1984. *Betalingsverkeer en officieuze economie in Nederland 1965–1982.* De Nederlandsche Bank n.v., monetaire monografieeen, 1.

Cramer JS. 1981a. The work money does: the transactions velocity of circulation of money in the Netherlands, 1950–1978. *European Economic Review,* 15: 307–326.

Cramer JS. 1981b. The volume of transactions and of payments in the United Kingdom 1968–1977. *Oxford Economic Papers,* N.S. 33: 234–255.

Cramer JS. 1986. The volume of transactions and the circulation of money in the United States 1950–1979. The Journal of Business and Economic Statistics, 4: 225–223.

Feige EL. 1979. How big is the irregular economy? *Challenge,* 22, November/December, 5–13.

Feige EL. 1980. *A new perspective on macroeconomic phenomena. The theory and measurement of the unobserved sector of the United States: causes, consequences and implications.* Presented at the American Economics Association Meetings, Denver.

Feige EL. 1985a. *The Swedish payments system and the underground economy.* Stockholm: Industrial Institute for Economic and Social Research.

Feige EL. 1985b. *The equation of exchange and national accounting systems.* International Conference on Income and Wealth, Noordwijk.

Feige EL. 1986. *Currency velocity and cash payments in the United States 1986.*

Fisher I. 1909. A new method of estimating the velocity of circulation of money. *Journal of the Royal Statistical Society,* 72.

Fisher I. 1911. *The purchasing power of money.* London: Macmillan.

Friedman M. 1956. The quantity theory of money – a restatement. *Studies in the Quantity Theory of Money,* Chicago: the University of Chicago Press.

Keynes JM. 1930. *A treatise on money.* London: Macmillan.

Keynes JM. 1936. *The general theory of employment, interest, and money.* London: Macmillan.

Keynes JM. 1956. *Essays and sketches in biography.* Meridian Books.

Laurent R. 1970. *Currency transfers by denomination.* Ph. D. Thesis, Chicago: University of Chicago Press.

Manski C and Goldin E. 1987. *The denomination-specific demand for currency in a high-inflation setting: the Israeli experience.* This volume.

Randall EB Jr and Mandel J. 1962. A statistical comparison of the wearing characteristics of two types of dollar notes. *Materials Research and Standards,* 2: 17–20.

Selden R. 1956. Monetary velocity in the United States, in: *Studies in the quantity theory of money,* edited by M Friedman. Chicago: University of Chicago Press.

Social and Cultural Report. 1980. Rijswijk: Social and Cultural Planning Office.

Spindt P et al. 1985. *A micro-analytic view of the payments mechanism: some preliminary results from the household survey of currency and transactions account usage.* Board of Governors of the Federal Reserve System.

The denomination-specific demand for currency in a high-inflation setting: the Israeli experience

CHARLES F. MANSKI and EPHRAIM GOLDIN

1. Introduction

The denominational mix of currency in circulation is part of the fine structure of a monetary economy. Although Chen (1976) has argued that the denominational mix has macroeconomic consequences, the prevailing view seems to be that the combination of banknotes and coins of different nominal values into the aggregate termed currency is a convenient, relatively harmless analytical simplification. This is evidenced by the fact that the denominational mix is seldom discussed in the monetary literature.

Whatever its proper role in monetary economics, the composition of the currency is a matter of great practical concern to central banks. By and large, central banks see their function in the currency realm as one of supplying the public demand. This means that the central bank should always have sufficient reserves to meet the demand for each of the denominations currently in use, as transmitted from the public by the commercial banking system. See, for example, Fase, van der Hoeven, and van Nieuwkerk (1979). Since currency inventories are costly to hold, it is important to the central bank to be able to forecast the denomination-specific currency demand. This need has led to a number of empirical investigations, including Fase (1981) in the Netherlands, Browne (1981) in Ireland, Kimball (1981) in the United States, and the present study in Israel.

The very high rate of inflation experienced in Israel has made the problem of forecasting the composition of currency demand particularly acute here. To understand the extent of the changes that have taken place over the past decade, consider the history of the ten-shekel (one hundred pound) banknote.[1] When this denomination was introduced in early 1969, its value in June 1981 prices was 710 Israel Shekels (IS), equivalent to about $62. By mid-1977, the value of the ten-shekel note had dropped to IS 130, or $11, in June 1981

[1] Introduced as a hundred-pound note, it became the ten-shekel note when the currency scale was divided by ten in 1979.

prices. At this point a new denomination, the fifty-shekel note, was introduced. In December 1980, when a hundred-shekel note was issued, the ten-shekel note was worth IS 14, or $1.20, in June 1981 prices. During 1982, the ten-shekel note was replaced by a coin.

The movements in the demand for the ten-shekel note since its introduction are summarized in Table 1. It is obvious that drastic changes have taken place. In particular, the peak demand occurred in 1976 and was eight times that of 1970 and double that of 1981. As a percentage of all notes in circulation, the ten-shekel denomination rose from 16 percent in 1970 to 50 percent in 1976 and then fell to 18 percent in 1981. In value terms, ten-shekel notes accounted for 46 percent of the note circulation in 1970, 84 percent in 1976 and only 4 percent in 1981. Because banknotes have consistently accounted for 97 to 98 percent of the value of all currency, the value shares of the ten-shekel note in the note circulation are very close to the corresponding shares in the currency circulation, not given in the table.[2]

Israel's high inflation rate and the associated large movements in currency

Table 1. Ten-shekel notes in circulation, 1969–1981.[a]

	Ten-shekel notes in circulation (millions)	Number of ten-shekel notes as percentage of number of all banknotes	Ten-shekel notes as percentage of value of all banknotes
1969	3	8	30
1970	5	16	46
1971	8	22	56
1972	12	27	64
1973	18	33	69
1974	23	39	76
1975	32	47	82
1976	41	50	84
1977	40	46	61
1978	35	39	38
1979	27	30	21
1980	23	21	9
1981	22	18	4

[a] Includes ten-shekel and hundred-pound notes. Circulation is defined as all notes not held by the Bank of Israel. Figures are for December 31 of each year.

[2] The definition of circulation used in Table 1 differs somewhat from the usual economic definition in that it includes notes held by the commercial banking system. In value terms, the commercial banks have tended to hold between 10 and 20 percent of the circulation, as defined here. The extended definition of circulation will be used throughout this paper as it is the one relevant to the central bank's denomination-specific demand forecasting problem.

demand greatly aggravate the Bank of Israel's task of planning note inventories but also provide a rare opportunity for research on the determinants of the composition of the currency demand. In particular, the Israeli data require a more structural approach than that found in Fase (1981), Browne (1981), and Kimball (1981).

The analyses of Fase, Browne, and Kimball were all performed in contexts of relatively slow inflation and fixed lists of currency denominations. The common approach of these authors was to estimate a system of note-demand equations, one for each denomination, the equations being related to one another only through a possible covariance of disturbances. The Fase, Browne, Kimball approach may produce adequate note demand forecasts in slow-inflation, fixed-denomination regimes. However, it cannot be satisfactory in a setting such as the Israeli one.

There are two reasons for this. First, the validity of a denomination-specific demand equation estimated from time series relies on the stability over time of the properties of the denomination. But if inflation is high, a crucial attribute of denomination, its real value, changes quickly. For example, it would make little sense to treat the ten-shekel note of 1969 and 1981 as the same good. Second, a system of denomination-specific equations provides little guidance in forecasting note demands when a new denomination is introduced. To make such forecasts, we must have a model which characterizes all potential denominations in terms of a common set of attributes, not one which treats denominations as qualitatively distinct goods.

To develop a model capable of forecasting currency demand in regimes with high inflation and changing denomination lists, we shall build on a suggestion of Cramer (1983). Cramer proposes deriving the composition of the currency demand from the distribution of real cash transactions and from the assumption that cash payments are made in the most efficient manner. An 'efficient' payment is one executed using denominations which minimize the number of pieces of paper and metal changing hands between buyer and seller.[3] From the distribution of transaction sizes and the assumption of efficient payment, one can derive the number of units of each denomination transacted, per unit time. Given additional assumptions relating the stock demand for a denomination to its flow use in exchange, the composition of the currency demand can be obtained.

In its pure form, Cramer's model is too idealized for empirical application. However, his general idea does offer the outline of a solution to the problem of forecasting note and coin demand in a rapidly changing environment. In particular, the Cramer proposal recognizes that the demand for a denomination depends not on its nominal value but on its real value relative to the

---

[3] The notion of efficient payment is also discussed in Chen (1976) and in Payne and Morgan (1981).

distribution of transactions. Moreover, the notion of efficient payment provides an operational way to characterize the degree of substitutability between notes of different denominations.

In what follows, Section 2 embeds the above ideas into an econometric model of the currency demand. Section 3 describes the Israeli currency system and the data used in the empirical analysis, presents parameter estimates, and analyses the findings.

2. A model of the currency demand

2.1 Definitions and assumptions

To introduce our model of the currency demand, we need first to define a number of terms:

- t — indexes time
- R — real value of a cash transaction
- $f_t(R)$ — density function expressing the distribution of cash transaction sizes, in real terms, at time t. By definition, $\int_0^\infty f_t(R)dR = 1$.
- K_t — number of cash transactions performed during period t.
- i — a banknote or coin denomination
- I_t — set of denominations offered at time t.
- D_i — nominal value of denomination i.
- P_t — price level at time t. By definition, D_i/P_t is the real value of denomination i at time t.
- $g_{it}(R)$ — function giving the expected number of units of denomination i used in a cash transaction of real size R, at time t.
- V_{it} — velocity of circulation of denomination i at time t.
- N_{it} — number of units of denomination i in the circulation at time t.

Using the above notation, we can write the identity

$$N_{it} \equiv \frac{K_t}{V_{it}} \int_0^\infty g_{it}(R) f_t(R) \, dR. \tag{1}$$

Equation (1) is the denomination-specific analogue to the familiar aggregate money identity $MV \equiv PT$. Equation (1) ceases to be an identity if we impose structure on its constituent elements, $f_t(\cdot)$, $g_{it}(\cdot)$, K_t, and V_{it}. This we now do.

It will be assumed that the distribution of real cash transaction sizes, $f_t(\cdot)$, is stable over time and has the form of a gamma distribution, so that

$$f_t(R) = \frac{\beta^\alpha}{\Gamma(\alpha)} R^{\alpha-1} e^{-\beta R}, \tag{2}$$

where $\alpha>0$ and $\beta>0$ are parameters of the distribution.[4] The family of gamma distributions is flexible enough to approximate well any actual transactions distribution, as long as that distribution is roughly single-peaked.[5] Some special cases are the family of exponential distributions ($\alpha = 1$, $\beta>0$) and the family of chi-square distributions ($\alpha>0$, $\beta = 1/2$).

Consider next the function $g_{it}(R)$. It is reasonable to suppose that the expected number of units of denomination i used in a transaction of real size R depends on two time-varying factors. One of these is the relationship between R and the real value, D_i/P_t, of the denomination. The other is the list of denominations $j \in I_t$, $j \neq i$, which could be used in place of or in combination with i to consummate the transaction. Thus, there should exist a time-invariant function $g_i(\cdot, \cdot, \cdot)$ such that

$$g_{it}(R) = g_i(R, \frac{D_i}{P_t}, I_t). \tag{3}$$

In the Cramer model, the function g_i is determined by the dictates of efficient payment. But the assumption that all payments are made efficiently seems too restrictive to be imposed *a priori* in empirical work. Given this, we have chosen to specify g using the flexible functional form

$$g_i(R, \frac{D_i}{P_t}, I_t) = c_{iI}\left(\frac{P_t R}{D_i}\right)^{a_{iI}} \exp\left(-b_{iI}\frac{P_t R}{D_i}\right), \tag{4}$$

where a_{iI}, B_{iI}, and c_{iI} are non-negative and $iI = (i, I_t)$. Equation (4) says that given any list of denominations I_t, the expected use of denomination i in transactions of real size R depends on the ratio of R to the real value of i. The way in which expected use depends on this ratio is determined by the three parameters (a, b, c), which are themselves functions of i and of the list of denominations. If $c>0$, $a>0$ and $b = 0$, equation (4) is a power function and use of denomination i is monotonic increasing in $P_t R/D_i$. If a, b, and c are all positive, equation (4) has a shape in the family of gamma densities, up to a scale factor.

When the structural assumptions (2) and (4) are imposed on the denomination-specific demand identity (1), we obtain

$$N_{it} = \frac{K_t}{V_{it}} \frac{\beta^\alpha c_{iI}}{\Gamma(\alpha)} \left(\frac{D_i}{P_t}\right)^{-a_{iI}} \{\int_0^\infty R^{(\alpha+a_{iI}-1)} \exp[-(\beta + b_{iI} P_t/D_i)R]dR\}. \tag{5}$$

Notice that the integrand in (5) has the form of a gamma density, up to a missing scale factor $(\beta + b_{iI} P_t/D_i)^{(\alpha+a_{iI})}/\Gamma(\alpha + a_{iI})$. It follows that the integral

---

[4] It can be questioned whether the distribution is stable over long periods. The assumption seems reasonable for the empirical work of this paper, which is based on data from 1969 to 1981.
[5] If prices are conventionally 'rounded off', the actual distribution may not be precisely single-peaked, but may have spikes at R values corresponding to round nominal values.

equals the reciprocal of this scale factor and therefore

$$N_{it} = \frac{K_t}{V_{it}} \cdot e^{\delta_{iI}} \left(\frac{D_i}{P_t}\right)^{-a_{iI}} (\beta + b_{iI} P_t/D_i)^{-(\alpha + a_{iI})}, \qquad (6)$$

where $e^{\delta_{iI}} = [\beta^\alpha c_{iI} \, \Gamma(\alpha + a_{iI})]/\Gamma(\alpha)$.

Thus, the demand for note or coin i at time t depends on the size of the cash economy K_t, on the denomination-specific, time-specific velocity of circulation V_{it}, on D_i/P_t, the real value of i, and on the list of denominations in circulation, I_t.

2.2 Econometric form of the model

The denomination-specific currency-demand function (6) may be written in logarithmic form as

$$\log N_{it} = \log K_t - \log V_{it} + \delta_{iI} - a_{iI} \log(D_i/P_t) - \\ - (\alpha + a_{iI}) \log(\beta + b_{iI} P_t/D_i). \qquad (7)$$

To obtain an estimable econometric model, we now modify (7) and impose additional assumptions.

First, consider K_t and V_{it}, which are not directly observable. Following conventional practice, we use the real gross national product as a proxy for K_t. Our handling of V_{it} requires more lengthy discussion.

It is reasonable to suppose that, all else equal, velocity of circulation increases with the opportunity cost of holding money. In the high inflation context of Israel, the opportunity cost C_{it} of holding a unit of denomination i in period t is well represented by

$$C_{it} = 100 \left(\frac{D_i}{P_{t-1}} - \frac{D_i}{P_t}\right) = 100 \left(\frac{P_t}{P_{t-1}} - 1\right) \frac{D_i}{P_t}, \qquad (8)$$

that is, by the percentage inflation rate multiplied by the unit's real value. A second possible determinant of velocity is the relative convenience of a denomination as a store of value. In particular, it is sometimes suggested that the highest-denomination note will be the favored instrument for hoarding and that, as a result, this denomination will have a lower velocity of circulation than do the smaller ones. We might therefore wish to let V_{it} depend on the position of i in the currency list.

In our empirical analysis, the above ideas are combined by letting

$$\log V_{it} = \gamma_1 C_{it} + \gamma_2 C_{it}^2 + \gamma_3 C_{it}^3 + U_{iI} \qquad (9)$$

where $\gamma_1, \gamma_2, \gamma_3$ are parameters and U_{iI} is a position-specific effect. The choice of a cubic polynomial in C allows us to model a fairly general relationship between opportunity cost and velocity. The term U_{iI} is not identifiable relative to δ_{iI} and so does not explicitly appear in the estimated model.

The second step in making equation (7) operational is to specify how the parameters δ_{iI}, a_{iI} and b_{iI} vary as functions of (i, I). A flexible specification is obtained by permitting δ_{iI}, a_{iI}, and b_{iI} to depend on the position of denomination i in the currency list I. That is, we differentiate between the highest, second-highest denomination and so on. Further, generality is obtained by allowing the level of note demand, represented by δ_{iI}, to depend on the distance between i and its closest substitute. In the Israeli context, this means distinguishing between denominations whose face values are powers of ten (i.e., 0.1, 1, 10, 100) and those which are five times powers of ten (i.e., 0.5, 5, 50, 500). Because the highest-denomination note has no substitute from above, we treat this case separately.[6]

A third step is to replace the non-linear term

$$-(\alpha + a_{iI}) \log(\beta + b_{iI} P_t/D_i)$$

by two linear-in-parameters approximations. To do this, observe that

$$-(\alpha + a_{iI}) \log(\beta + b_{iI} P_t/D_i) = -(\alpha + a_{iI}) \log \beta -$$
$$- (\alpha + a_{iI}) \log\left(1 + \frac{b_{iI}}{\beta} \frac{P_t}{D_i}\right). \quad (10)$$

If for given (i, I), $(b_{iI}/\beta)(P_t/D_i)$ is far from zero for all t, then $\log[1 + (b_{iI}/\beta)(P_t/D_i)] \simeq \log(b_{iI}/\beta) - \log(D_i/P_t)$, so that (10) reduces to $-(\alpha + a_{iI}) \log b_{iI} + (\alpha + a_{iI}) \log(D_i/P_t)$. Returning to (7), the first term of the above expression can be incorporated into δ_{iI} and the second term can be combined with the term $-a_{iI} \log(D_i/P_t)$, yielding $\alpha \log(D_i/P_t)$. If $(b_{iI}/\beta)(P_t/D_i)$ is close to zero for all t, the first order approximation

$$\log\left(1 + \frac{b_{iI}}{\beta} \frac{P_t}{D_i}\right) \simeq \frac{b_{iI}}{\beta} \frac{P_t}{D_i}$$

is valid and (10) reduces to

$$-(\alpha + a_{iI}) \log \beta - (\alpha + a_{iI}) \frac{b_{iI}}{\beta} \frac{P_t}{D_i}$$

To handle this case, we need only introduce P_t/D_i as an explanatory variable, whose coefficient depends on the position of denomination i in the currency ladder. Since we do not know whether $(b_{iI}/\beta)(P_t/D_i)$ is close to or far from zero, we include P_t/D_i in all situations. The only situation which may not be satisfactorily covered by these approximations is one in which $(b_{iI}/\beta)(P_t/D_i)$ is

---

[6] We do not let a_{iI} and b_{iI} depend on the distance to the nearest substitute, as there is little reason to think that any relationship should exist.

close to zero in early periods but, as the price level rises, becomes much greater than zero in later periods of the sample data.

Our fourth modification to (7) is to allow for the possibility that the number of units of denomination i in circulation differs from the demand because of lags in adjustment to changing circumstances. Postulating a conventional stock-adjustment mechanism, we add the lagged endogenous variable log $N_{i)t-1)}$ to the list of explanatory variables. Special treatment of adjustment is necessary when a new denomination is introduced or an old one withdrawn from circulation. The estimated model incorporates additional variables to deal with these situations.

To complete the econometric version of the currency demand model, we add a disturbance ε_{it} to equation (7). We permit the disturbances to be first-order serially correlated and the estimation procedure takes this possibility into account. On the other hand, we ignore any correlation that may exist between denominations.[7] Model estimation is via a three-stage process. In the first stage log N_{it} is regressed on all exogenous variables and on their values lagged one period. In the second stage, the predicted values of log $N_{i(t-1)}$ from the first-stage regression are used as instruments for the lagged endogenous variables log $N_{i(t-1)}$ and an ordinary-least-squares instrumental-variables estimate of the structural model is obtained. The residuals from this second-stage regression are then used to estimate the serial-correlation coefficient of the disturbances. The third stage re-estimates the structural model, now by generalized least squares. Under the assumptions of the model, the third-stage parameter estimates are consistent, asymptotically normal, and asymptotically efficient. If, contrary to the assumption, the disturbances are correlated across denominations, the estimates are not asymptotically efficient but remain consistent and asymptotically normal.

3. Empirical analysis

3.1 Data

The estimates to be presented are based on monthly data for the period January 1969 through March 1981.[8] Monthly denomination-specific series on currency circulation were available for banknotes but not for coins. For this

---

[7] Estimation of a model whose disturbances are both serially correlated and correlated across denominations would be very complex, particularly as the inter-denominational correlation structure is not likely to be stable.

[8] The exception was the GNP series, available quarterly. A monthly series was obtained by simply assuming GNP to be constant within each quarter.

reason, we use note observations only.[9] The number of such observations was 641, there co-existing four or five note denominations in each of the 147 months of the sample. The evolution of the list of note denominations over the sample period is described in Table 2.

Descriptive statistics for the variables appearing in the estimated model are given in Table 3. This table also provides the precise definitions of the variables as measured.

3.2 Findings

Examination of the parameter estimates in Table 4 reveals that three factors – real note value, position in the currency list, and opportunity cost-explain most of the substantial observed variation in note demand between denominations and over time. In this section, we analyse the influence of each of these major determinants of note demand and discuss the roles of other, empirically less important factors. Following the component-by-component analysis, we assess the overall fit of the model.

3.3 Real value of notes

We find that for the highest denomination, the quantity of notes in circulation

Table 2. Note denominations in circulation.

	IS 0.1 (IL 1)	IS 0.5 (IL 5)	IS 1 (IL 10)	IS 5 (IL 50)	IS 10 (IL 100)	IS 50 (IL 500)	IS 100
1969	√	√	√	√	√		
1970		√	√	√	√		
1971		√	√	√	√		
1972		√	√	√	√		
1973		√	√	√	√		
1974		√	√	√	√		
1975		√	√	√	√		
1976		√	√	√	√		
1977		√	√	√	√	√	
1978		√	√	√	√	√	
1979		√	√	√	√	√	
1980			√	√	√	√	
1981			√	√	√	√	√

[9] Even if denomination-specific coin circulation series were available, their use would problematic. Because Israeli coins have often had very low real values, public handling of coins is thought to be careless, leading to large losses. Thus, the effective coin circulation may be noticeably smaller than the official circulation.

increases sharply as real value declines, while for the other denominations, the relationship between real value and note demand is less pronounced. The position-specific effects of real value are most easily seen by reference to Fig. 1.

In each graph, the horizontal axis gives the approximate range of real values

Table 3. Descriptive sample statistics.

Variable	Mean	Standard deviation
log note volume[a] (log N_{it})	2.53	0.68
Dummies H_i	0.228	0.420
H_1: for highest denomination note		
H_2: for second-highest note	0.228	0.420
H_3: for third-highest note	0.228	0.420
H_4: for fourth-highest note	0.228	0.420
Dummy for power-of-ten face value and $H_1 = 0$	0.319	0.467
log of real note value[b]	0.01	0.28
$H_1 \log(D_i/P_t)$		
$H_2 \log(D_i/P_t)$	−0.21	0.54
$H_3 \log(D_i/P_t)$	−0.52	0.99
$H_4 \log(D_i/P_t)$	−0.73	1.40
$H_5 \log(D_i/P_t)$	−0.42	1.35
Reciprocal of real note value	0.26	0.61
$H_1 (P_t/D_i)$		
$H_2 (P_t/D_i)$	0.83	2.38
$H_3 (P_t/D_i)$	2.63	6.14
$H_4 (P_t/D_i)$	8.22	23.79
$H_5 (P_t/D_i)$	12.33	49.53
Opportunity cost (C_{it})	0.82	1.179
Opportunity cost squared (C_{it}^2)	2.06	5.73
Opportunity cost cubed (C_{it}^3)	7.86	33.0
log real GNP[c]	−0.858	0.187
Linear time trend (t)	76.9	43.3
Dummy for period 1/69–4/73	0.335	0.472
Dummy for period 5/73–10/77	0.346	0.476
New-note variable[d]	0.021	0.118
Withdrawn-note variable[e]	0.746	3.446

[a] Note volume is in millions of units (Bank of Israel figures for the last day of each month).
[b] Here and elsewhere, the price level is measured relative to that of September 1951, for which $P_t = 1$. The term 'price level' refers to the Consumer Price Index.
[c] Real GNP is nominal GNP, in IS millions per year, divided by the price level. The nominal GNP series is available quarterly and is assumed constant within each quarter.
[d] Letting s(i) be the date at which denomination i is introduced into circulation, the variable takes the value $1 - [t - s(i)]/12$ for $0 \leq t - s(i) \leq 12$ and zero thereafter.
[e] Letting u(i) be the date at which denomination i is last supplied to the public by the Bank of Israel, the variable takes the value $t - u(i)$ for $0 \leq t - u(i) \leq 24$. For $t - u(i) > 24$, we consider the note fully withdrawn from circulation.

observed during the sample period. For example, the lowest observed real value for a highest-denomination note is 0.23, which was the value (in terms of the shekel of year 1951) of the fifty-shekel note in late 1980, just before introduction of the hundred-shekel note.[10] The vertical axis gives the predicted ratio of note demand at a given real value to that at the lowest observed real value. It is important to point out that the plotted curves express the transactions effect of real note value, operating through the variables log(D/P) and

Table 4. Estimated note-demand model.

Variable	Coefficient	Asymptotic standard error
Constant	1.68	0.91
Dummies H_i	1.46	0.96
H_1: for highest denomination note		
H_2: for second-highest note	0.37	0.95
H_3: for third-highest note	0.45	1.00
H_4: for fourth-highest note	−0.67	1.02
Dummy for power-of-ten face value and $H_1 = 0$	0.81	0.03
log of real note value	−1.83	0.09
$H_1 \log(D_i/P_t)$		
$H_2 \log(D_i/P_t)$	0.11	0.05
$H_3 \log(D_i/P_t)$	0.42	0.08
$H_4 \log(D_i/P_t)$	−0.13	0.05
$H_5 \log(D_i/P_t)$	0.02	0.11
Reciprocal of real note value	−0.8505	0.0538
$H_1 (P_t/D_i)$		
$H_2 (P_t/D_i)$	−0.0122	0.0070
$H_3 (P_t/D_i)$	0.0231	0.0046
$H_4 (P_t/D_i)$	0.0013	0.0007
$H_5 (P_t/D_i)$	0.0022	0.0007
Opportunity cost (C_{it})	0.042	0.028
Opportunity cost squared (C_{it}^2)	−0.034	0.0101
Opportunity cost cubed (C_{it}^3)	0.00325	0.00096
log real GNP	−0.0046	0.0366
Linear time trend (t)	0.0040	0.0005
Dummy for period 1/69–4/73	0.05	0.04
Dummy for period 5/73–10/77	0.06	0.02
Lagged dependent variable	0.099	0.014
New-note variable	−1.20	0.06
Withdrawn-note variable	−0.056	0.003
First-order serial-correlation coefficient	0.83	0.02
R^2 of the third stage regression	0.95	
Sample size	641	

[10] A real value of 0.23 corresponds to about $6.40 in June 1981 dollars.

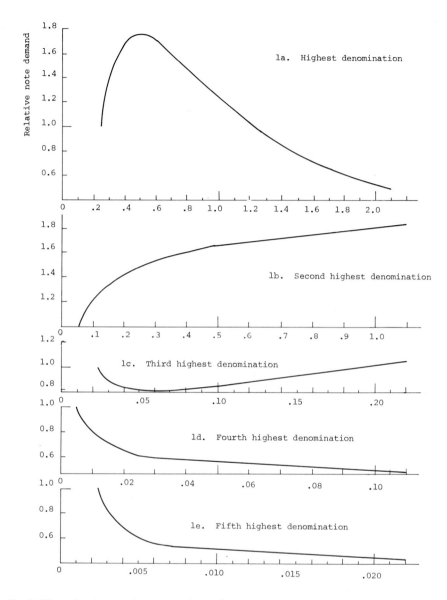

Fig. 1. Effect of real note value on note demand.

P/D explaining log N. The conceptually distinct influence of real value through its effect on opportunity cost will be discussed separately.

Why should the graph for the highest-denomination note be different from the others? A simple and persuasive explanation emerges from the idea of efficient payment. Let the denominations I be ordered in terms of decreasing value, so that i = 1 is the highest denomination note, and consider a transaction of real size R. If $R<D_{(i+1)}/P_t$, denomination i will be rarely used, it being more efficient to use denomination i + 1.[11] For transactions in the range $D_{(i+1)}/P_t<R<D_i/P_t$, use of note i will become increasingly frequent. For $R>D_i/P_t$, the pattern depends crucially on whether i is the highest-denomination note. If i = 1, we should expect that use of the note will grow more or less proportionately with R. In fact, efficient payment would require that the number of units of i used in a transaction of size R be the integer closest to P_tR/D_i, denoted $[P_tR/D_i]$. On the other hand, if i>1, the existence of a higher denomination i − 1 puts a ceiling on the number of units of i needed for efficient payment. We should therefore expect to see that the pattern established in the range $0 \leq R \leq D_{(i-1)}/P_t$ will be repeated cyclically in each of the ranges $\lambda D_{(i-1)}/P_t \leq R \leq (\lambda + 1) D_{(i-1)}/P_t$, $\lambda = 1, 2, 3, \ldots, \infty$.

With this as background, consider the effect of an increase in the price level, yielding a decrease in real note values. One result is that for each denomination i, the fraction of transactions satisfying the condition $R>D_{(i+1)}/P_{(t+1)}$ is larger than the fraction for which $R>D_{(i+1)}/P_t$. Thus, an increase in the price level should yield an increase in the number of transactions involving at least one unit of note i and, thereby, an increase in the demand for i. For i>1, there is no further systematic effect, an increase in P being equivalent to a decrease in the cycle index λ. However, for i = 1, a second important effect does operate. That is, for transactions of size $R>D_i/P_t$, the number of units of i needed to make efficient payment increases from $[P_tR/D_i]$ to $[P_{(t+1)}R/D_i]$. It follows that an increase in the price level should lead to a more than proportional increase in the demand for the highest denomination note and to a smaller increase in the demand for other denominations.

In most respects, our empirical results are consistent with the above reasoning. As Fig. 1 shows, the demand for the highest denomination is much more sensitive to real note value than are the demands for other notes. Moreover, the parameter estimates of Table 4 imply that the real-value elasticity of demand for the highest denomination is $-1.83 + 0.85(D_i/P_t)^{-1}$, which is more negative than −1 for all D/P>1.02. Of course, inspection of Figure 1 makes it clear that the theory of efficient payment cannot explain all aspects of our empirical findings. In particular, the demand for the second-highest denomination is noticeably increasing rather than decreasing as a function of real note value. In addition, the elasticity of demand for the highest-denomination note approaches zero for low real note values. At extremely low real values the elasticity even turns positive. This last phenomenon is somewhat mysterious as the data do not exhibit such a reversal. The phenomenon is presumably an artifact of our functional form specification. Within the sample, only ten or so observations lie within the positive elasticity region.

[11] For the lowest denomination note, denomination i + 1 is a coin.

3.4 Position in the currency list

Let us now turn to the effect of position *per se*. The coefficients of the position-specific dummies H_1–H_4 each capture a variety of effects and hence are not easily interpretable. Of greater interest is the dummy variable distinguishing between notes i whose face values are powers of ten and those whose face values are five times powers of ten, given that i>1, Our statistically very precise parameter estimate, 0.81, implies that all else equal, demand for a denomination of the former kind is $e^{0.81} = 2.25$ times as high as demand for one of the latter kind.

It seems clear that this phenomenon derives from the different degrees of substitutability of the two kinds of notes. If denomination i is a power of ten, then i − 1 has five times the value of i and efficient payment may involve use of as many as three units of i in some transactions. On the other hand, if i is five times a power of ten, then i − 1 has double the value of i and efficient payment never requires more than one unit of i. In light of this, our estimate that demand in the first case is 2.25 times that in the second seems very plausible. It also seems reasonable to speculate that if two-shekel and twenty-shekel notes were to be added to the existing list (1, 5, 10, 50, and 100 shekels), face value might cease to be an important determinant of note demand.

3.5 Opportunity cost

In equation (8), we defined the opportunity cost of holding a unit of denomination i during period t to be $C_{it} = 100(P_t/P_{t-1} - 1)(D_i/P_t)$. However, this function expresses realized, not expected, opportunity costs and the latter seems the more relevant concept here. Preliminary empirical investigations led us to conclude that expected inflation in period t is well modelled by the six-months geometric average inflation $(P_t/P_{t-6})^{1/6}$. Therefore, the opportunity cost variables appearing in Table 4 are defined as $C_{it} = 100[(P_t/P_{t-6})^{1/6} - 1](D_i/P_t)$ and this quantity squared and cubed.

We have employed a cubic polynomial in C_{it} in an attempt to discover the nuances of the relationship between opportunity cost and note demand. Our findings are most easily seen by reference to Fig. 2. The figure shows that when C_{it} is close to zero, increases in C_{it} have little or no effect on note demand. As C_{it} moves away from zero, a negative relationship begins to develop and grows in intensity. However, for high values of C_{it}, this negative relationship attenuates until, at the extreme right-hand limit of the sample range, further increases in C_{it} yield no further decreases in demand. Overall, the effect of an increase in opportunity cost from the left-hand sample limit, $C_{it} = 0$, to the right-hand limit, $C_{it} = 6.5$, is to reduce note demand by 25 percent.

Our finding that note demand is a monotonic decreasing function of oppor-

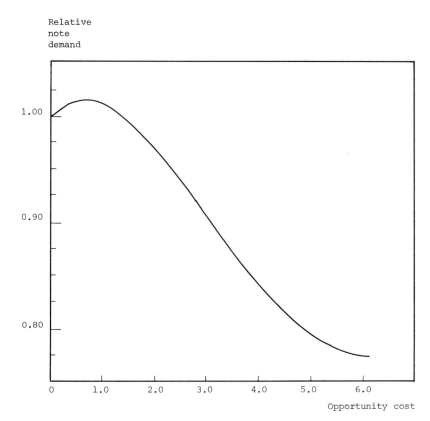

Fig. 2. Effect of opportunity cost on note demand.

tunity cost is fully expected. Perhaps less expected is the attenuation of the cost-demand relationship at high cost values. We should point out that the predicted total flattening out of this relationship may be an artifact of the cubic specification. However, there is reason to believe that some attenuation really does occur, as a consequence of the increasing difficulty of consummating transactions as the quantity of currency in circulation falls. This matter deserves further exploration.

Although our variable C provides a theoretically appealing operational definition of opportunity cost, it may be asked how alternative specifications perform. One possibility is to let the inflation rate interact with position in the currency list rather than multiply it by real note value. We explored specifications of this type and found that the rate of inflation has a noticeable influence only on the demand for the highest denomination note. This result is consistent with our opportunity-cost findings.

3.6 Time-varying variables

Table 4 indicates the absence of any discernible linear relationship between log real GNP and log N. An alternative specification, using real private expenditure as the measure of the size of the cash economy, fared no better. Suspecting that our disaggregation of currency into its components might somehow mask the size effect, we experimented with conventional aggregate currency-demand specifications as well. We still found no GNP or private expenditure effect.

In a further attempt to see whether note demand changes systematically over time, we introduced two period-specific dummy variables and a linear time trend. The former proved impotent but the latter uncovered a small, but statistically very significant positive trend. It appears that all else equal, note demand is increasing at the rate of 0.4 percent per month, or 4.9 percent per year. This value is of the expected order of magnitude for a size effect and we are tempted to interpret the time trend as such. However, the success of the trend variable relative to the more explicit GNP and private expenditure variables remains unexplained.

3.7 Adjustment variables

The quantity of notes in circulation apparently adjusts quite quickly to changes in demand. Our parameter estimate for the lagged dependent variables is 0.099, implying that 90 percent of the adjustment to an exogenous shock occurs in the first month. This rate of adjustment is very much faster than that found in aggregate currency equations estimated in other countries (see, for example, Goldfeld, 1973). It is natural to ask whether our result indicates exceptionally speedy adjustment in Israel or whether it is related to our disaggregation of the currency into denominations. To investigate this, we estimated a conventional aggregate-demand model and obtained a parameter value of 0.50 for the period up to 1977 and 0.20 thereafter. These figures are higher than that in the denomination-specific model but are still much lower than in other aggregate-currency models. It thus appears that the speed of currency adjustment in Israel really is high.

When a new note is introduced into circulation, the adjustment process is much lengthier, presumably because of the need for the public to become accustomed to the new denomination. Our estimated coefficient for the new-note variable implies that a new note attains $e^{-1.20} = 0.30$ of its equilibrium circulation in the first month after introduction and $e^{-1.20/2} = 0.55$ after six months.

When a decision is made to withdraw a note from circulation, the Bank of Israel stops supplying notes to the public and removes from circulation all notes deposited with it. Otherwise, the Bank allows notes in the hands of the public to remain in use. Our estimated coefficient for the withdrawn-note variable implies that 71 percent of the old notes are still in circulation six months after the formal withdrawal date, 51 percent after a year, and 26 percent after two years. The withdrawal process is thus relatively slow.

3.8 Serial correlation of the disturbances

As is common with monthly data, the disturbances of the note-demand model exhibit strong serial correlation. The estimated first-order serial correlation coefficient, obtained from the second-stage, instrumental-variables regression residuals, is 0.83 and is very significant statistically. This estimate is used as an input into the third stage, approximate generalized least-squares regression. With the exception of the serial correlation coefficient, all estimates in Table 4 are from the third-stage regression.

3.9 Fit of the estimated model to the data

In Figs 3a to 3e, we plot for each denomination the actual number of notes in circulation and our prediction of note demand, based on the estimated model. The plotted prediction for denomination i in period t is the value

$$\hat{N}_{it} = \left(\exp \frac{X_{it}\hat{\beta}}{1-\hat{\lambda}}\right). \tag{11}$$

Here $\hat{\lambda} = 0.099$ is the estimated coefficient of the lagged dependent variable, taken from Table 4. The vector of explanatory variables exclusive of the lagged dependent variable is X_{it} and $\hat{\beta}$ is the corresponding vector of parameter estimates. Equation (11) is the natural formula for long-run prediction of note demand and provides a much more stringent test of the model's predictive ability than would a formula for short-run prediction.[12]

The diagrams show that for the lower denominations, the in-sample fit of the model is exceptional. For the half-shekel denomination, predictions track realizations almost perfectly. For the one-shekel note, the fit is very close except in the final months of the sample period during which the predictions are considerably above the realizations. These deviations are presumably due to the fact that since late 1980, the one-shekel note has co-existed with a one shekel coin. In the case of the five-shekel denomination, the predictions track

---

[12] Equation (11) is not the best formula for short-run predictions because it does not use information contained in the adjustment process and in the serial correlation of the disturbances. Figure 3 does not include plots for the one-pound and one-hundred-shekel denominations because these denominations were available in note form only at the very beginning and end of the sample period, respectively.

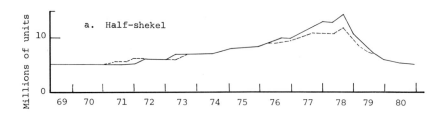
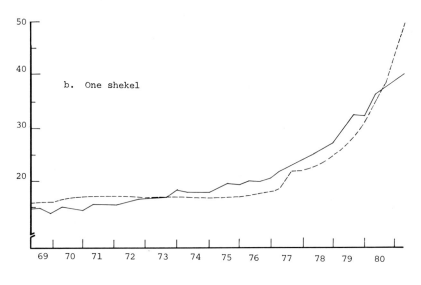
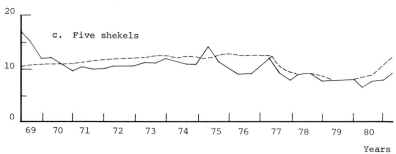

Fig. 3. Note demand by denomination: actual and predicted.

——— Actual

-------- Predicted

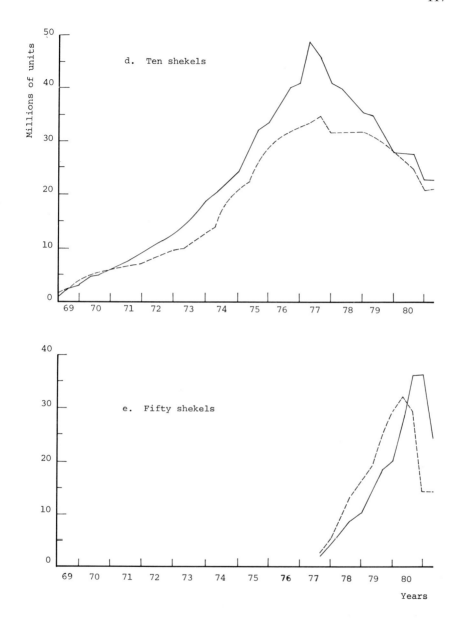

Fig. 3. Note demand by denomination: actual and predicted.

———— Actual

-------- Predicted

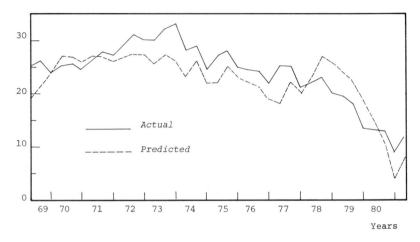

Fig. 4. Real-note demand: actual and predicted.

realizations well except at the beginning of the sample period. At this time, the five-shekel note was in transition from being the highest denomination in circulation to being second highest.

Consider now the ten-shekel and fifty-shekel denominations. These deserve special attention as the former was the highest note in circulation from 1969 to early 1977 and the latter from early 1977 through 1980. Inspection of Figure 3d reveals that for the ten-shekel note, predictions are close to realizations except in 1976 and 1977, during which predictions are significantly lower than realizations. For the fifty-shekel note, the fit is reasonably good until early 1980, at which point the predictions turn downward prematurely. These phenomena almost certainly have a common cause. In both cases, the deviations occur when the highest-denomination note has its lowest real value. But this is exactly the regime within which the estimated real-value elasticity of note demand turns positive (see above, the section on the real value of notes). We therefore caution against use of the model in this regime.

While our concern in this paper has been to explain the denomination-specific currency demand, it might be asked whether the estimated model can be used to predict the aggregate real demand for currency. Evidence that the model performs fairly well in this application is provided by Fig. 4, which plots the actual real-note circulation and our prediction, defined as

$$\hat{M}_t = \sum_{i=1}^{I_t} N_{it}\left(\frac{D_i}{P_t}\right). \tag{12}$$

We have made no attempt to compare the quality of the predictions in Figure 4 with what can be obtained using a conventional aggregate-currency equation.

4. Conclusions

The findings presented here explain much of the observed pattern of Israeli currency demand, between denominations and over time. During the sample period, the Israeli economy experienced drastic movements in denomination-specific currency demands and changes in the currency list. Nevertheless, the estimated model fits the historical data quite well. It is impressive that the statistical success of the model is achieved using a small number of explanatory factors – the real value of a denomination, its position in the currency list, and the opportunity cost of holding a note. Moreover, the manner in which these factors are found to act on the currency demand is consistent, for the most part, with a set of simple structural hypotheses. These are that the flow use of currency is determined by the need to consummate transactions, that payments are made efficiently, and that the velocity of circulation of a denomination depends on the opportunity cost of holding it.

5. Acknowledgement

This research was performed as part of a project carried out at the Currency Department, Bank of Israel. We are grateful to Shmuel Peled, Head of the Currency Department, for the opportunity to conduct the research and for much constructive feedback during its execution. At the Bank, Shmuel Aviezer, Aaron Gilshon, Itzhak Na'aman, and Shlomo Sirkis contributed useful suggestions at various stages of the project. Ariel Guttman assisted ably in the data collection. We would like to thank J.S. Cramer for a number of stimulating discussions on the subject of currency demand.

Bibliography

Browne FX. 1981. Forecasting the demand for currency and its denominational mix. Technical Paper 5/RT/81, Central Bank of Ireland, Dublin.

Chen C. 1976. Currency denominations and the price level. *Journal of Political Economy*, 81, 1: 179–183.

Cramer JS. 1983. Currency by denominations. *Economics Letters*, 12, 299–303.

Fase MMG, Hoeven D van der and Nieuwkerk W van. 1979. A numerical planning model for a central bank's bank note operations. *Statistica Neerlandica*, 33, 1: 7–25.

Fase MMG. 1981. Forecasting the demand for banknotes: some empirical results for the Netherlands. *European Journal of Operational Research*, 6, 3: 269–278.
Goldfeld S. 1973. The demand for money revisited. *Brookings Papers on Economic Activity*, 3: 577–638.
Kimball RC. 1981. Currency only part of money supply. *Coin World*, 11: 81.
Payne LC and Morgan HM. 1981. U.K. currency needs in the 1980s. *The Banker*, April 1981, 45–53.

In search of 100 billion dollars

MARIUS VAN NIEUWKERK

1. Introduction

Cramer prefers to explore unmapped territory, doing research that presents all the challenge and excitement of discovery. From estimators, parameters and distributions to motor-cars and bank notes. Unfettered by convention and inspired by a high degree of originality. Doubly fascinated when the search is difficult, tracing that rock-solid parameter or that cast-iron constant or those many 'missing' Dutch 1000-guilder bank notes.

This contribution also deals with 'missing' money. Not guilders this time but dollars. Not in the Netherlands but in the world: the statistical discrepancy in the world balance of payments.

2. The world statistical discrepancy

It all seems so very simple. Exports from one country are imports into another. Exports from all countries combined equal imports into all these countries. Hence, by definition world exports or world imports equal world trade.

Let us take this a bit further: the difference between a country's exports and its imports gives rise to a net balance in the form of a surplus or a deficit. A surplus in one country means a deficit in another. The total of all surpluses and deficits in the world should be nil.

That is how simple it should be; yet reality is different. If for the year 1984, for instance, the current-account surpluses and deficits (on account of visible trade, services, factor income and transfers) of all countries are totalled, the result is a world deficit of nearly 100 billion dollars. This means that world payments for goods, services and income exceed world receipts by this amount. So, something somewhere must be wrong.

This 'statistical discrepancy' in world current-account balances is, for several reasons, a major source of concern for both statisticians and policy makers. First, it is clear that the statistics are inadequate. The crucial question is, however, in what countries and in what respects. Second, the statistical discrepancy has increased sharply in recent years. In 1976 it totalled only 6 billion dollars and in 1977–1979 it averaged some 20 billion dollars a year. The period since 1980, however, has seen a sharp increase and in the past few years the amount has fluctuated around 100 billion dollars, thereby seriously aggravating the problem.

Under these conditions, policy makers are on increasingly slippery ground. Are the amounts recorded for receipts too low or do recorded payments exceed actual payments? In other words: what is the level of world trade? Are the surpluses of some countries in fact higher and/or are the deficits of other countries lower than the amounts officially recorded? Could it be that, as a result, surplus countries do too little to stimulate their economies whereas deficit countries place undue emphasis on austerity policies? Phrased differently, does the negative statistical discrepancy in the world balance of payments mean that the world economy is swathed in a deflationary mist?

As yet, there are no definite answers to these questions. To end this unsatisfactory situation, the IMF instituted a Working Party of international experts in early 1985; they were given two years to resolve the problems. As a member of this Working Party, I shall now attempt to unveil part of the mystery.

3. The world current account

The world current account is compiled by the IMF by totalling the balance-of-payments data reported by the member countries (cf. Table 1). On the basis of the available estimates, it is assumed that the lack of information about some countries which do not report to the IMF (mainly Eastern bloc countries) is not the principal cause of the ever larger discrepancy.

The world current account is divided into the same categories as the national current accounts: goods, services and income, etc. It is noteworthy that the various components show not only negative discrepancies (mainly Services and Official transfers) but positive discrepancies as well (Trade and Private transfers). Apparently, the overall discrepancy is due to various factors. Another striking feature is that some components are marked by a more or less stable discrepancy, whereas the discrepancies for other components show wide fluctuations and/or sharp increases. The latter is notably true of a number of categories of the Service balance. Especially investment income from bank deposits, securities, etc. has shown a large negative discrepancy in recent years; see Other investment income (item 2f). Apparently, in these cases reported receipts have been increasingly lower than reported payments.

4. The discrepancy

Considering the relatively short period of time available to resolve the problems, eight meetings in two years, the Working Party opted for a clear-cut and low-cost strategy. This strategy aims at utilizing to the full all supplementary sources available to the IMF itself. In the event that these supplementary sources should prove insufficient, additional specific information will have to be sought, such as data by country, by financial centre or by transport centre.

The first question raised by the Working Party was to what extent reported transactions are classified incorrectly by the national compilers. A case in point is the positive discrepancy in the Trade balance, where receipts from exports include transport earnings which should have been classified in the transport balance; in the transport balance this has led to negative discrepancies. Similar shifts may occur between reinvested earnings and distributed earnings in respect of direct investment. Moreover, official transfers may have been recorded as private transfers by recipient countries.

These are serious discrepancies, each of which is given due attention by the Working Party. By utilizing available sources and by means of supplementary surveys, information has been sought about such items as transport margins (differences between c.i.f. and f.o.b. values) and Private and Official transfers. The resulting reclassification will probably be made largely *within* the

Table 1. Selected balances of world current-account transactions (In billions of US-dollars).

			1978	1980	1982	1983	1984
1.	Trade balance		18.1	28.2	−2.0	9.8	11.0
2.	Service balance		−24.7	−49.2	−100.9	−78.7	−96.4
	a)	Shipment	−24.2	−32.0	−33.8	−31.8	−33.5
	b)	Other transportation	−1.7	−3.4	−4.4	−3.4	−1.1
	c)	Travel	−0.3	−0.9	1.5	3.2	4.5
	d)	Reinvested earnings on direct investments	6.7	11.2	7.5	9.9	5.8
	e)	Other direct investment income	−4.6	−7.6	−11.3	−11.5	−11.7
	f)	Other investment income	−6.2	−11.2	−35.9	−32.0	−41.6
	g)	Other official transactions	−4.0	−11.4	−24.0	−18.2	−20.5
	h)	Other private transactions	9.6	6.2	−0.4	5.1	1.8
3.	Private transfers		4.5	7.0	3.8	6.7	3.7
4.	Official transfers		−17.5	−20.8	−14.8	−12.9	−14.2
5.	Current account (1 through 4)		−19.7	−34.7	−113.9	−75.1	−95.8
	(Memo: Service balance as a percent of service payments)		(5.8)	(7.1)	(12.8)	(10.9)	(12.7)

world current account. Hence, it is not expected that it is these discrepancies which have underlain the increasing overall discrepancy on the world current account.

Subsequently, the Working Party looked into the effects of transactions which are not reported fully, either because one country reports them whereas another does not, or because they are only partially reported by a country. A good example is provided by capital flight from country A to country B.

Let us assume that this is done by smuggling gold from country A to country B, the gold being sold in the latter country and the proceeds being placed on deposit with a local bank.

Table 2. Different discrepancies in the world balance of payments (Fictitious amounts).

	Country A			Country B			World (A+B)		
	Credits	Debits	Net	Credits	Debits	Net	Credits	Debits	Net
1. *Adequate reporting*									
Trade balance	100		100		100	−100	100	100	0
Capital balance		100	−100	100		100	100	100	0
Total	100	100	0	100	100	0	200	200	0
2. *Non-reporting country A*									
Trade balance	0		0		100	−100	0	100	−100
Capital balance		0	0	100		100	100	0	100
Total	0	0	0	100	100	0	100	100	0
3. *Imperfect reporting countries A and B*									
Trade balance	80		80		100	−100	80	100	−20
Capital balance		100	−100	90		90	90	100	−10
Total	80	100	−20*	90	100	−10*	170	200	−30*
4. *Non-reporting of both countries*									
Trade balance	0		0		0	0	0	0	0
Capital balance		0	0	0		0	0	0	0
Total	0	0	0	0	0	0	0	0	0

* Errors and omissions (E & O's).

If everything is reported by the book, the balance-of-payments entries are as shown in Table 2 under 1. A trade surplus in country A of a (fictitious) amount of 100 is offset by an equal deficit in country B. The world trade balance is in equilibrium. The same is true of the capital balance: country A exports capital (by increasing its assets) while country B imports capital to the same amount. Both national balances of payments and the world balance of payments are in equilibrium; each entry is offset by a contra-entry and there are no residual items (errors and omissions).

However, let us suppose that, contrary to country B, country A does *not* record these transactions at all in its balance of payments. In that case, even though the three balances of payments continue to have no residual items, the components of the world balance of payments will be marked by opposite discrepancies: a deficit on current account and a surplus on capital account (Table 2, under 2). If the ensuing transactions continue to be unreported, the investment income earned by, and the resulting new capital outflows from, country A will lead to similar discrepancies, albeit that the negative discrepancy on current account now arises in the income balance (Other investment income).

A much more difficult situation is created if the transactions are not reported correctly, so that the balances of payments do not add up and errors and omissions arise (Table 2 under 3). Operating as it does from IMF headquarters, the Working Party cannot do much to reduce these errors and omissions, as they may be distributed among a number of countries and may concern various balance-of-payments items. The reparation of these errors and omissions is primarily the responsibility of the national compilers. In this case, the world balance of payments offers few clues for the search to eliminate the discrepancies.

In the third place, the Working Party was aware of the trap that would be set if neither of the two countries were to report their reciprocal transactions. Even though this would seem to solve the problem because there would be neither discrepancies nor errors and omissions, it is clear that this would be utterly unacceptable (Table 2 under 4). Additional information must then be obtained to ascertain to what extent receipts and payments have been fully reported (cf. Section 5).

In order to obtain an impression of the degree to which, in the world balance of payments, the discrepancies on current account are reflected in opposite discrepancies on capital account, world totals have been calculated for these two accounts. Table 3 shows that until 1977 the cumulated discrepancies were almost negligible. However, for the period 1977–1983 the cumulated discrepancies on the world current account were nearly 350 billion dollars. Their counterpart was to be found to about two-thirds in the capital account and to one-third in errors and omissions.

This result provided the Working Party with a significant clue, indicating as it did that various countries systematically fail to record both receipts on current account and payments on capital account (cf. Table 2 under 2). This meant that the many data on capital transactions available from other statistics might be used to advantage in analysing the discrepancies. Examples are notably financial stock data relating to the various assets and liabilities of countries, which are abundantly available in the form of banking and other financial statistics.

5. Other investment income

One category from the world current account which lends itself eminently to such analysis is Other investment income (income from portfolio investments and loans, excluding income from direct investment). Increasing sevenfold from 6 billion dollars in 1978 to 42 billion dollars in 1984, the discrepancy in this item has been the fastest growing and most significant within the current account (Table 1, item 2f). This means that reported payments (income debits) have been consistently higher than reported receipts (credits), so that most probably receipts have been underreported.

Thus far, the IMF did little to relate the results of the balance-of-payments reporting to the data available from other sources. Consequently, the Working Party first of all tried to establish a connection between the capital *flow* data from the balance-of-payments statistics and the *changes* in the relevant financial stocks evident from other statistics. About these stocks, a wealth of reliable supplementary statistics is available, such as the international assets and liabilities statistics of the banking system (reporting to the BIS) and of the private and government sectors (reporting to the OECD and the IMF).

Table 3. Main sectors of world balance-of-payments accounts (In billions of US-dollars; debits (-)).

	Cumulated 1964–76	Cumulated 1977–83
Current account	−38	−347
Capital movements	34	237
of which: increase of liabilities	(892)	(2,670)
increase of assets	(−858)	(−2,433)
Errors and omissions	4	111
of which: credit entries	(34)	(285)
debit entries	(−30)	(−174)

The aim is to compile a consistent overall survey of financial stocks and flows, broken down geographically. Once such a survey is available, the stock data can also be used to assess the plausibility of investment income as included in the balances of payments, since income must be in a certain proportion (rate of return) to stocks. For many countries this proportion proved to be far from plausible. Consequently, a selection was made of countries with a high standard of reporting and plausible rates of return. Using these data and supplementary market information, calculations were made to determine plausible rates of return at the world level. Multiplying these rates by the relevant stocks produces estimates of investment income. Table 4 shows the results of such calculations for the year 1983.

The average rate of return comes out at 9%. Hence, the income flow generated by the $4,360 billion of stocks should be $380 billion. The actual amounts reported were about $325 billion in income credits and about $355 billion in income debits. Therefore, the adjustments to be made are $55 billion and $25 billion respectively. So, the non-reported share of credits is some 15% and that of debits 6–7%; the difference between the two shares might be due to the fact that, for mainly tax reasons, interest received from portfolio investment and loans tends to be reported to a lesser extent than such interest paid.

Thus the net amount to be adjusted is about $30 billion, which is of course roughly equal to the discrepancy. This amount has been broken down geographically in Table 5.

The largest single adjustment is accounted for by the group of Industrial countries. Within this group, it is especially the United States for which considerable adjustments need to be made; with regard to non-bank claims on

Table 4. Estimate of investment income on world cross-border financial positions, 1983 (excluding direct investment).

Positions	Outstanding amounts $ billion	Rate of return percent	Investment income $ billion
Interbank accounts	1,900	9.5	176
Bank claims on nonbanks	724	10	71
Nonbank claims on banks	631	9	54
Foreign official reserves	182	10	18
Claims of international organizations	140	7	10
Bonds	400	9.7	36
Equities	150	4	6
Inter-nonbank positions	235	4	9
Total	4,360	9	380

banks alone, this country accounted for some 60% of non-reported interest earnings. The rest of the amount was distributed fairly equally among the other industrial countries. The non-reported amount of interest earnings from bonds accounted for by this group of industrial countries constituted a substantial part (80%) of the non-reported amount at the world level. With regard to these results, it must be noted that residents from these industrial countries may also hold funds temporarily in off-shore centres, as the ultimate beneficial owners of these funds are to be found throughout the world. This means that the results for the countries other than off-shore centres shown in Table 5 represent minima.

This might also explain why such surprisingly small adjustments are needed for the oil-producing countries in the Middle East. The adjustments required for the other developing countries are much larger (nearly half relating to Latin America). Here, too, part of the off-shore adjustment is probably ultimately accounted for by this group of countries. Finally, part of the adjustment remains geographically unallocated, because not all the data reported are sufficiently broken down by country. This, too, means that the eventual adjustments would turn out to be larger.

6. What causes the discrepancies

The question what causes the growing discrepancies is of even greater importance than the effort to eliminate them. To be able to answer this question, we must take a look at developments in the highly-developed industrial countries. If only because of their share in the world economy and international finance, these countries must inevitably account for a major part of the discrepancies. Another consideration is that developments in these countries are often ahead of those in other countries, so that they may afford an enlightening example.

Table 5. Allocation of non-direct investment income discrepancy by areas (In billions of US-dollars, net amounts).

Country/groups	Reported data	Adjustments	Adjusted data
Industrial countries	−6.6	15.5	8.9
Middle East oil exporting countries	24.8	1.1	25.9
Major off-shore centres	0.6	5.7	6.3
Other developing countries	−50.7	6.4	−44.3
Other and unallocated	–	3.7	3.7
Total (see Table 1, line 2f)	−31.9	32.4	0.5

In the first place, it must be noted that, from the early 1980s onwards, deregulation has been the word in these countries. On the one hand, this means that policy is increasingly oriented towards relieving industry of official paperwork, while, on the other, the financial means of the statistical services were curbed, and their manpower curtailed. This undoubtedly led to a decline in the quality of the statistics.

Secondly, these countries are the first to introduce all sorts of new financial instruments and new financial paper. The considerable flexibility of these innovations goes hand in hand with decreased possibilities for proper statistical registration.

In the third place the registration of balance-of-payments statistics is often undertaken by a variety of institutions in these countries, especially the United States and the United Kingdom. Furthermore, the surveys conducted are not always attuned to one another, which means, in a world marked by deregulation and financial innovations, an extra reduction of the possibilities for a proper registration of the statistics.

The fourth factor is inherent in the growing liberalization in these countries, namely the increasing use made of tax havens. Discrepancies can arise from capital flight or tax-routing from one industrial country to another, via these tax havens. A striking example is that where the investments of country H (Head) are not made directly in country T (Tail), but run via one or more off-shore centres. That means that instead of the direct flows:

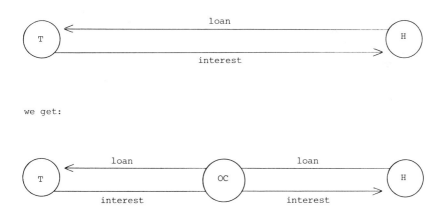

As can be seen from the adjustments made with regard to Other investment income, these off-shore centres are by no means neutral channels. Either they transform direct investment flows into portfolio investment flows, receiving on balance income from portfolio investments and paying direct investment income, or the income from portfolio investments is hoarded and reinvested.

It goes without saying that both this income and these capital flows are relevant for the world balance of payments.

7. Conclusion

During the past decade the world may have become more liberalized, it has certainly not become easier for statisticians and researchers to monitor. In his study of the holdership of liquidity in the Netherlands, Cramer (1973) already concluded that, especially with regard to the missing bank notes, 'it might be useful to investigate whether other forms of surveys would not lead to better results'. This is even more so where registration on a global scale is concerned. Although definitive conclusions in respect of the statistical discrepancy in the world balance of payments would be premature, one thing has become clear. Under the present, changed circumstances, the compilation of satisfactory balance-of-payments statistics on the basis of one single source and/or by means of one single registration method no longer suffices, not only with regard to statistics on a national level, but also in respect of those of international institutions. The Working Party will consequently put forward various recommendations on how to make better use of sources from outside the balance-of-payments sphere. Proposals will also be made about how to achieve closer cooperation between statisticians, researchers and policy makers in the compilation of statistics.

One thing is clear, and that is that in this day and age quite a few financial transactions have come to be hidden from sight. As the situation is increasingly getting out of hand, expectations are that counter-forces will have to be mobilized, especially from the viewpoint of policy-making. In addition to the above proposals for a better use of the existing sources of statistics and other data there will probably be increasing calls for measures ensuring a better international exchange of information. Rules will have to be laid down, for example, about who is resident where and who is required to report to which authorities. For the time being we will have to make do with inadequate statistics and analyses. That means, on the one hand, greater discrepancies and uncertainty, and, on the other – luckily for them – a greater challenge and more excitement for economic researchers.

Bibliography

Cramer JS and Reekers GM. 1973. *Het houderschap van liquiditeiten in Nederland.* Amsterdam: NIBE.

Forecasting the daily balance of the Dutch Giro

AART F. DE VOS

1. Introduction

This paper is based on a consultancy project for the Dutch Postal Clearing Service, the 'Giro'. The Giro plays an important part in transactions in the Netherlands. Specifically salary payments to households are important. As most of these take place on rather fixed days each month, and also have yearly patterns, the balance of the giro shows strong calendar variations and seasonality. Apart from that there are clearly trend movements. To develop a model to forecast all these movements was the goal of the project. This succeeded but the resulting model appeared much more complex than expected. Instead of the balance several flows of money going into and out of the system were modelled. Moreover each flow was decomposed into monthly aggregates showing trends and seasonality and the distribution of the aggregates over the days in each month showing all kinds of calendar effects. For all the submodels several possibilities have been tried, evolving from ad hoc solutions based on traditional time-series models to solutions based on the Kalman-filter. Around 1983 we discovered the latter possibilities, mainly due to the work of Harvey (1981). The Kalman-filter has a great appeal as a unified framework. Moreover the certainty that the specified models are optimally estimated – often not possible with ad hoc solutions – is reassuring.

The difference between the ad hoc solutions and the elegant models proved much smaller however than we had hoped. We spent some time looking for the causes and the consequences. The only thing we will mention at the end of the article is about the consequences: it appears to be possible to estimate the quality of a model that combines the best of both models in it. The task remains to develop such a model.

This paper is an essay in honour of Cramer. He taught me econometrics, he encouraged the solving of real problems with much attention for the data and respect for, but no worshipping of, techniques. Rereading his introduction to his 'empiricial econometrics' (1969) for this occasion I realized how far his view

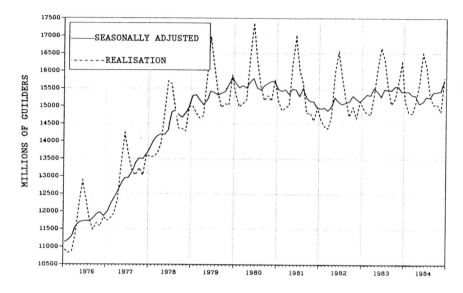

Fig. 1. Giro Balance (monthly figures).

on econometrics has penetrated my approach of problems. His introduction stresses the compatibility of empirical evidence with a wide range of alternative hypotheses; the irrelevance of the statistical theory of testing in econometric work; the requirement of understanding of what is relevant to the particular observations at hand. His spirit should be recognizable in this paper.

2. The Giro balance and its short- and long-term fluctuations

Figure 1 gives the development of the balance of the Dutch Giro (Postal Cheque and Clearing Service, after 1-1-1986 demand deposits of the Postbank). With a level since 1980 of approximately 15 billions of guilders this is about 20% of M_1 in the Netherlands. For private households however, its importance is much higher. Also the other way round: private accounts are a very important part of the balance. Fluctuations are mainly caused by monthly salary payments. In May when holiday allowances are paid, the balance reaches a peak. A smaller peak is reached in December. The seasonal pattern is very regular: seasonal adjustment in Graph 1 is almost equal to adjustment of the logarithm of the series by a fixed pattern estimated by zero sum dummies. (In fact the method of Harvey and Todd (1983) to be treated later on is used which allows changing seasonality, but the change is very small). The trend movement is mainly due to the development of national income. As reported in De Vos (1983) a unit elasticity is acceptable, and the effect of interest movements is negligible. Finally the movements within the month: these are also very regular with as main effect a rise around the end of the month when most salaries are paid and a decrease during the rest of the month. Calendar effects are important however: the situation of the weekends and holidays matters of course but also the length of the month and many smaller items.

The regularity of the monthly and yearly patterns suggests that there are many ways to reproduce these. In fact two methods, based on exponential smoothing techniques, had been used before the project started. The reasons to look for something better were twofold:
- The rather abrupt turn of the trend development at the end of 1978 was only after a long time captured by the models, so considerable overestimation of the long-term development took place.
- Short term forecasts, important for cash-management decisions had absolute errors of over 100 million and their preparation was very labor-intensive.

These are rather diverse features. In fact short-term and long-term forecasting are almost independent activities. Though it is possible in theory that a better model for daily fluctuations improves the recognition of the trend movements, in practice this effect is negligible. In our models we used a decomposition into monthly aggregates and daily fluctuations in these aggregates. This decomposition allows to separate the modelling of short-run and long-run features.

We confined ourselves to time-series models. The only proper exogenous variable is national income. We did not succeed thusfar in using this variable as a proper tool in forecasting. Monthly national income figures are not there, the quarterly figures are only available some months after the quarter is over. It is quite likely that the development of the giro balance itself is a more up-to-date indicator of the development of national income. In that case this exogeneous variable is of no use in forecasting.

3. Balance, inflow and outflow

We chose not to model the balance itself but the money flows into and out of the system; the connection being the identity

$$S_t = S_{t-1} + I_t - U_t \tag{1}$$

with S_t the balance at the end of day t, and I_t the inflow and U_t the outflow during day t. This setup has several advantages. The inflow, I_t, consists for a large part of salary payments which have a very specific calendar structure. Moreover in some cases there exists exogenous information on salary payments. Strongly this advantage was felt when in 1985 holiday allowances for government employees were largely paid in March instead of May. Government payments and some other important inflows are modelled separately, so this kind of information can easily be dealt with. Money-flows are also a more natural entity than balances, especially when monthly aggregates and the distribution over the month are considered. The monthly aggregates will appear to depend on the number of trading-days in the month, a feature that may be used in modelling the distribution. With money balances this kind of consistency is much more difficult to reach. The outflow is treated differently. It seems natural to explain outflow at least partly from the balance. We spent some time looking for a good specification for the dependence of S_t on past outflows. For weekly data a remarkable regression result emerged:

$$\begin{aligned} S_w = 0.74I_w &+ 0.66I_{w-1} + 0.74I_{w-2} + 0.69I_{w-3} + 0.59I_{w-4} \\ &+ 0.58I_{w-5} + 0.36I_{w-6} + 0.21I_{w-7} + 0.13I_{w-8} + 0.20I_{w-9} \\ &+ 0.20I_{w-10} + 0.14I_{w-11} + 0.09I_{w-12} + 0.23I_{w-13} + 0.21I_{w-14} \\ &+ 0.15I_{w-15} + 0.14I_{w-16} + 0.19I_{w-17} + 0.11I_{w-18} + 0.08I_{w-19} \end{aligned} \quad (2)$$

With standard errors of about 0.07 and residual autocorrelation ϱ_k of approximately $(.92)^k$. As the coefficients of (2) are about $.75\,(.92)^k$, (2) is very strong evidence that

$$S_w = .92 S_{w-1} + .75 I_w \quad (2a)$$

holds. For daily figures the relation

$$S_t = .973 S_{t-1} + I_t + \varepsilon_t \quad (3)$$

arose, so using (1)

$$U_t = .027 S_{t-1} + \varepsilon_t \quad (4)$$

We used this result as a starting point for a new approach: U_t/S_{t-1}, the outflow percentage, as dependent variable in a time series model describing trends, seasonality and calendar effects in a similar fashion as I_t. The subsequent developments proved that we did the right thing: between 1979 and 1982 a structural shift in the outflow percentage took place due to a structural shift in payment habits (see Fig. 2). A model using a fixed coefficient – statistically in 1980 a justifiable choice – would have led to systematic biases in the forecasts.

The procedure to forecast the balance has become quite complicated by all these decompositions. Series for several inflows and the outflow percentage and decomposed into monthly aggregates and the distributions over the month; these aspects are forecast with time series methods (to be treated later on) and finally all parts are taken together to get a forecast of the balance. All this works, but whether it works under all circumstances better than forecasting the balance itself is not clear. For daily forecasts the decomposition is justified as the in- and outflow series behave rather differently. For long-term forecasts the gain should come from the relative stability of the outflow percentage: if this is the case then shifts in the inflow have a predictable influence on the balance. When every coefficient changes stochastically however, it becomes very complicated to get a clear picture of gains and losses of several models.

4. Trends and seasonality

Monthly aggregated inflows and monthly means of outflow percentages are modelled as:

$$Y_m = \ln(z_m/(\alpha + (1-\alpha)\frac{R_m}{21})) = d_m + s_m + u_m \tag{5}$$

with

z_m	= the dependent variables
R_m	= number of trading days (mean 21)
$0<\alpha<1$	= coefficient describing dependence on R_m
d_m	= trend
s_m	= seasonal component
u_m	= disturbance term.

Several possibilities have been tried for the specification of d_m, s_m and u_m. The logarithmic specification is chosen after some experiments with the Box-Cox transformation (advocated in De Vos (1976, 1982) primarily to get a seasonal pattern with constant amplitude). The specification of the influence of R_m is chosen in a way that allows consistent modelling of calendar effects. R_m however is a crude measure of the influence of the calendar, it may later be replaced by a variable derived from the calendar models. The estimates of α do not depend strongly on the specification of d and s. They are rather diverse for several series. Inflow series have a value for α of about .5: half of the inflow is monthly bound, the other half daily. The outflow however, is only for 11% monthly bound. This stands to reason: salary payments are a main source of inflow, and these are monthly bound.

The specification of trends and seasonality is a well-known subject where many confounding discussions were held about. The early literature was dominated by non-statistical algorithms that accomplished a decomposition of a series into trend, seasonal and irregular. The Census X-11 method (see Young (1968)) is the most famous of these and still in use at many places. Comparisons between different algorithms failed however, due to lack of clear objectives (De Vos (1975)). Subsequent trials to use statistical models have been hampered by the dominance of the multiplicative seasonal ARIMA models of Box and Jenkins (1970). The most common form of this model is:

$$(1-L)(1-L^s)y_t = (1-\theta L)(1-\theta L^s)\varepsilon_t \qquad (6)$$

with s the number of seasons (12 when monthly figures), L the lag-operator and ε_t white noise.

A drawback of this specification is that it allows no natural decomposition into trend and season. Also, the parameters do not correspond with concepts like the variability of season and trend. Models that do have these characteristics are called 'structural models' by Harvey (1984). In Harvey and Todd (1983) a model – proposed earlier by Harrison and Stevens (1976) within a Bayesian framework – is used:

$$\begin{aligned} y_t &= d_t + s_t + u_t \\ d_t &= d_{t-1} + g_t + \varepsilon_t \\ g_t &= g_{t-1} + \eta_t \\ s_t &= -\sum_{i=1}^{s-1} s_{t-i} + \zeta_t \end{aligned} \qquad (7)$$

with u_t, ε_t, η_t and ζ_t independent variates, having variances σ_u^2, σ_ε^2, σ_η^2 and σ_ζ^2.

The main problem with this model within the Box-Jenkins framework, especially for monthly data, is that (7) is equivalent to a model like (6), but with s + 1 parameters at the right hand side, depending in a nonlinear way on the relevant 3 parameters in (7), being $\sigma_\varepsilon^2/\sigma_u^2$, σ_η^2/σ_u^2 and $\sigma_\zeta^2/\sigma_u^2$. A rather complicated problem results described in De Vos (1982). In De Vos (1976) a model similar to (7) was proposed, but estimated with fixed seasonal dummies; in that case standard time series methods may be used. The first model we developed for the monthly aggregates was basically model (7), but estimated by an approximation from De Vos (1982). It is not worthwhile to give this here as the Kalman filter solution of the estimation problem is straightforward. Harvey and Todd (1983) and Harvey (1984) give the details. We will not repeat them here; when we treat the models for calendar variation we will give classical as well as Kalman-filter solutions worked out fully.

It is a standard result within the Kalman filter that $\sigma_\varepsilon^2/\sigma_u^2$, σ_η^2/σ_u^2 and $\sigma_\zeta^2/\sigma_u^2$ are the parameters. These parameters correspond with the variability in level and slope of the trend and the variability of the seasonal component. Full information maximum likelihood extimators may numerically be derived. Conditional upon these estimates all components may be estimated, including the standard errors. Moreover, this is possible with information up to the time an observation got available and with all observations (using the 'smoothing algorithm'). See Harvey (1981, ch. 4) for a general treatment. Still we did finally not use model (6) but a model proposed by Gersch and Kitagawa (1983), which in the same notation is:

$$y_t = d_t + s_t + p_t + u_t$$
$$d_t = 2d_{t-1} - d_{t-2} + \eta_t$$
$$p_t = \varrho p_{t-1} + \delta_t \qquad (8)$$
$$s_t = -\sum_{i=1}^{s-1} s_{t-i} + \zeta_t$$

The differences between (7) and (8) are very subtle. Basically (8) supposes that a really smooth trend exists, with a stationary first order autocorrelation process superimposed on it. For ϱ approaching unity model (8) approaches model (7), but none of our series gave estimates of ϱ higher than .8. Still the empirical results are almost equal. Only the estimated components differ $d_t + p_t$ (not smoothed) from (8) hardly differs from d_t from (7) except in interpretation. $d_t - d_{t-1}$ from (8) is almost equal to g_t in (7). Apparently our concept of trend is not yet clearly defined. What we probably have in mind when talking about trends is the ex post, smoothed, estimator. In this respect (8) gives what we want: the ex post trend (d_t) is in all cases a really smooth line. Fig. 2 gives an example. The meaning of the ex post trend should not be exaggerated; the most recent values may change drastically when new figures become available. But it is remarkable that a good fitting and forecasting model defines unambiguously such a trend.

The reason that we prefer (8) above (7) is not primarily the beautiful pictures. The main reason is the possibility to estimate the variance σ_η^2 that fixes the variability of the slope, which is very important for long term forecasts. In model (7) this variance is in many cases hardly estimable. In Harvey and Todd (1983) most empirical examples end up with $\sigma_\gamma^2 = 0$, which implies a constant slope of rather deterministic drift in d_t. This possibility is not surprising as ε_t in (7) may take care of structural shifts. In model (8) $\sigma_\eta^2 = 0$ is impossible, at least it would imply a deterministic linear trend. So it is not surprising that we find well-behaved estimates of σ_η^2 using model (8). Besides we found that for a number of series the forecasts 12 month ahead with model (8) were decidedly better than with model (7), a finding that is also reported by Gersch and Kitagawa (1983). Methodologically this is a remarkable feature: likelihood and most other criteria only look at forecasts one period ahead, but apparently there may be more things of importance.

5. Models for calendar effects

The calendar is an interesting phenomenon crowded with exogenous variables that may perfectly be forecast. This is a very important advantage in forecasting. The calendar is also a complex phenomenon: months have different lengths and weeks interfere with them. Yearly effects are not easily incorpor-

Fig. 2. Outflow percentage (indexed).

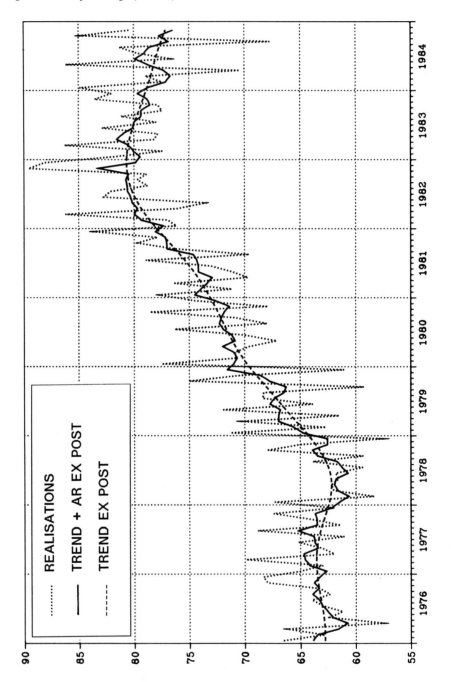

ated. The literature on calendar effects is rather meagre and in most cases more concerned with calendar effects on monthly aggregates than with the daily figures themselves.

Most models assume, at least partly, deterministic effects described by dummy variables. This is, especially in financial series, not very plausible. Structural shifts in calendar effects are likely as institutional changes occur frequently.

The recent literature contains some examples. Bell and Hillmer (1983) use a deterministic model for trading day effects on monthly aggregates. Pierce, Grupe and Cleveland (1984) give a model for weekly aggregates, deterministic again and estimated in a confusing way using spectral analysis; next they use these aggregates to model stochastic monthly effects. Pfefferman and Fisher (1982) give a procedure that is much like our basic adjustment for the number of working days given at the beginning of § 4. We will give the two models we constructed for the daily series that are part of the model. First the somewhat ad hoc solution to get the calendar in a more or less standard time series form, next a solution based on Kalman filters. It will not be possible to treat all details.

6. The Procrustes transformation

The first solution we chose for the calendar problem was a rather rigorous one: we transformed the calendar, or rather the division of the monthly aggregates. The histogram of the daily money flows (closing days and weekends excluded) is divided into 21 parts, 21 being the mean number of opening days. As this resembles shortening months that are too long and lengthening months that are too short, we called the transformation after Procrustes who in Greek mythology applied similar transformations to human beings. A simple example: to come from 5 to 4 days the transformation is:

1	1/4	0	0	0
0	3/4	1/2	0	0
0	0	1/2	3/4	0
0	0	0	1/4	1

The transformed data are used to forecast transformed monthly figures that are in a similar way 'laid back in the calendar'. From 4 to 5 days would imply the backtransformation:

4/5	0	0	0
1/5	3/5	0	0
0	2/5	2/5	0
0	0	3/5	1/5
0	0	0	4/5

It should be noted that these transformations do not fulfill the requirement that a pattern after transformation and backtransformation has remained the same. In fact the pattern is smoothed somewhat. In practice this effect appeared to be negligible.

The transformed data refer to 12 months of 21 'days'. This offers the opportunity to specify stochastically changing monthly and yearly patterns.

The weekly patterns and effects of holidays must however be described by a deterministic model to keep things manageable. To eliminate these effects while keeping the monthly and yearly patterns intact we need the model

$$Py_t = PD\alpha + u_i$$
$$u_i = u_{i-252} + \varepsilon_i \tag{9}$$

Here y_t is the original series. P is the Procrustes transformation (transformed series having i as index). D is a set of dummies: a constant term, four dummies for weekeffects (summing to zero) and some dummies for holidays. u_i is supposed to follow a 'random walk yearly pattern' which is easily implemented as each year has 252 'days'. The regression estimate of α according to (9) is straightforward. Though the regressor, yearly differences of Procrustes-transformed weekly dummies, is quite complicated no problem arises. Worth mentioning is the constant term that after the Procrustes transformation is equal to $R_t/21$, with R_t the number of opening days in the month. The coefficient of this 'constant' must in view of the model for the monthly aggregates (5) be (approximately) equal to $1-\alpha$, leaving a fraction α for the monthly bound effects.

The first change in the specification of the yearly pattern was to suppose for u_i a random walk plus disturbance term:

$$u_i = y_i + \zeta_i$$
$$y_i = y_{i-252} + \eta_i \tag{10}$$

an easy way to implement this model and moreover to get estimates of the pattern is the algorithm

$$y_i = y_{i-252} + \alpha\varepsilon_{i-252}$$
$$\varepsilon_i = u_i - y_i \tag{11}$$

(11) is equivalent to (10). This follows simply by rewriting (10) as:

$$u_i - u_{i-252} = \eta_i + \zeta_i - \zeta_{i-252} \tag{10a}$$

and (11) as

$$u_i - u_{i-252} = \varepsilon_i - (1-\alpha)\varepsilon_{i-252}. \tag{11a}$$

the right hand sides of (10a) and (11a) both only have a yearly, negative autocovariance. Equating these and the variances show that α is a function of $\alpha_\zeta^2 \alpha_\eta^2$. The form (11) is very easy to estimate, using the 'back-forecasting' algorithm of Box and Jenkins. The simplicity and elegance of this algorithm makes us repeat it here:

Start with equating y_i to u_i in the first year (necessary as y_i is a nonstationary process). Next use (11) for given α to go through the series and predict the postsample year. Then turn round: use the predictions as starting values and go backwards through the sample. 'Predict' the presample year. Turn again: use the presample year as starting values and now use (11) to get values of ε_i ($i = 0, \ldots$). $\sum \varepsilon_i^2$ is the criterion in the (numerical) optimization of α.

The simplicity of this procedure may be a reason to formulate models in the form (11) rather than (10). In more complicated cases this may be a way to avoid expensive estimation procedures. This is relevant when there are many data, as is the case for daily figures. So we did, we extended (11) to:

$$\begin{aligned} y_i &= y_{i-252} + \alpha\varepsilon_{i-252} \\ m_i &= m_{i-21} + \beta\varepsilon_{i-21} \\ \varepsilon_i &= u_i - y_i - m_i \end{aligned} \tag{12}$$

Now α and β are estimated optimizing $\sum \varepsilon^2$ as before. This proved to be a clear improvement for almost all series: a reduction of about 10% of the standard error of forecasts. There are some funny aspects of model (12). First it only uses one degree of freedom more than model (11): the 21 extra starting values for m play no role. A change in starting values for m compensated by a change in the starting values for y (for each change in one m twelve corresponding y's are changed) changes all values of m and y, generated by the algorithm, in the same way while the ε's remain unchanged. A unique definition of y and u would require restrictions. If only forecasts are required this is not necessary.

A second point to be noted about (12) is that it does not correspond with the 'structural' model

$$\begin{aligned} y_i &= y_{i-252} + \eta_i \\ m_i &= m_{i-21} + \zeta_i \\ u_i &= y_i + m_i + \varepsilon_i \end{aligned} \quad \text{with } \eta_i, \zeta_i, \varepsilon_i \text{ independent variates.} \tag{13}$$

To compare (12) and (13) it is convenient to use the fact that both models may be seen as 21 independent processes that follow some kind of 'random walk plus seasonality' pattern. For corresponding days (we use an index j to denote this) we may write (12) as

$$y_j = y_{j-12} + \alpha\varepsilon_{j-12}$$
$$m_j = m_{j-1} + \beta\varepsilon_{j-1} \qquad (14)$$
$$\varepsilon_j = u_j - y_j - m_j$$

By taking the twelfth difference of the third equation and substitution of the first equations we obtain

$$u_j - u_{j-12} = \varepsilon_j + \beta \sum_{k=1}^{11} \varepsilon_{j-k} + (\alpha + \beta - 1)\varepsilon_{j-12} \qquad (15)$$

while (13) in the notation like (14) would imply

$$u_j - u_{j-12} = \varepsilon_j - \varepsilon_{j-12} + \sum_{k=0}^{11} \zeta_{j-k} + \eta_j. \qquad (16)$$

Working out the variances and covariances of the right hand sides shows that the structure is similar, but not equal. Whether this is sufficient reason to prefer (12) is open to debate. Our experience is that models with similar characteristics give almost equal results and our feeling is that all models are only approximations, so we have no problems in using (12).

The differences between (12) and (11) are small however. The forecast errors have a correlation of .9. This confirms that the only difference is one degree of freedom. The gain by model (12) lies in a quicker adaption to changes in monthly patterns.

7. A Kalman-filter model for calendar effects

Not only models for trend and seasonality may easily be implemented with help of the Kalman-filter, also calendar models. A very important point is that the filter may easily deal with unequally spaced occurrence of a phenomenon, like the last day of the month. A state vector of calendar effects is just adapted dependent on what effects are relevant for each measurement. Even the Procrustes transformation could be cast in Kalman-filter form, the state vector being the 'transformed calendar'. The models we worked with are of the form:

$$y_t = \alpha_{wt} + \beta_{mt} + \gamma_{yt} + u_t \qquad (17)$$

$$\text{dummy}_t = \text{last corresponding dummy} + \eta_t \qquad (18)$$

α_w being five weekly dummies, β_m monthly dummies, and γ_y yearly dummies. As to the monthly dummies a wide variety of specifications is possible. As monthly bound payments must take place every month this is a restriction on the possible dummies too. The 21st of the month cannot be used: it may be in a weekend. We experimented with some specifications (using fixed dummies)

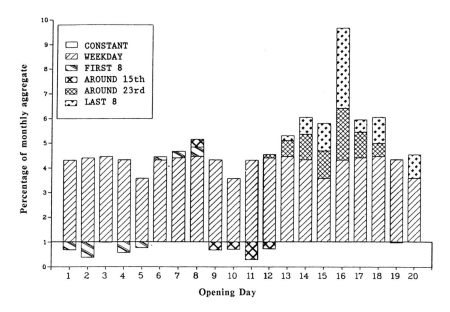

Fig. 3. Calandar effects of an inflow component.

and chose the first and last eight opening days, five days around the 15th (i.e. the 15th or if that is in a weekend the day before that weekend together with the two opening days before and after this date) and seven days around the 23rd (same construction), altogether 28 monthly dummies. Days around the 27th and first Monday of the month etc. had no significant influence.

For the yearly effect the same sets of dummies were tried, but bound to a specific month. Only for December seven days around the 19th and the last five opening days of the year survived the tests. Finally two dummies for days after holidays and special closing days made the total number of dummies 47.

Figure 3 illustrates the working of the model in a standard month for one of the inflow categories. The 'constant' is the part of the monthly bound money flow that is not incorporated in the monthly dummies; on a daily base it is $c + 21/R_t$. The day-bound constant effect is incorporated in the weekly dummies.

To estimate the model (17)–(18) it is necessary to specify restrictions on the variances of η_t. We assumed equal variances within the three groups: one for weekly, one for monthly and one for yearly dummies. These variances – or rather their ratio's to σ_m^2 – are estimated; the estimated values define how quickly the dummies are updated.

We will give the filter equations in the notation of Harvey (1981, ch. 4). The form is a specific case of the general framework. The measurement equation is a rewritten version of (17):

$$y_t = z'_t a_t + \xi_t \quad (19)$$

with y_t the vector of observations at the state vector containing all (47) calendar effects considered. (5 weekdays, 28 monthly effects and 14 yearly effects). z'_t contains a 1 at places corresponding to dummies that apply at time t and 0 elsewhere.

The transition equation, denoting how the state vector changes at time t, is simple due to the fact that the transition matrix is the unit matrix (it is e.g. (u)).

$$a_t = a_{t-1} + R_t \eta_t \quad (20)$$

with $Q = E(\eta_t \eta'_t) = \begin{bmatrix} \sigma_w^2 & 0 & 0 \\ 0 & \sigma_m^2 & 0 \\ 0 & 0 & \sigma_y^2 \end{bmatrix}$

R_t selects the variates that are relevant at time t: it has three columns with in the first column a 1 at the place of the relevant weekly effect, in the second column a 1 at the relevant monthly effect(s), in the third at the yearly effect(s) that apply and zeros elsewhere.

The prediction equations are simply

$$a_{t/t-1} = a_{t-1}; \quad y_{t/t-1} = z'_t a_{t-1} \quad (21)$$

and

$$P_{t/t-1} = P_{t-1} + R_t Q R'_t \quad (22)$$

And the updating equations

$$a_t = a_{t-1} + P_{t/t-1} z_t (y_t - z'_t a_{t-1})/f_t \quad (23)$$

$$P_t = P_{t/t-1} - P_{t/t-1} z_t z'_t P_{t/t-1}/f_t \quad (24)$$

$$f_t = z'_t P_{t/t-1} z_t + 1 \quad (25)$$

Finally the function to be optimized is the loglikelihood, following from the fact that the forecast errors $y_t - z'_t a_{t-1}$ are independent with variance $\sigma^2 f_t$. This implies that

$$\hat{\sigma}^2 = \frac{1}{n} \sum (y_t - z'_t a_{t-1})/f_t \quad (26)$$

To estimate the model σ_w^2, σ_m^2 and σ_η^2 are varied numerically. For P_0 a diagonal matrix with large values on the diagonal is chosen and the algorithm is used to get the forecast errors

$$v_t = y_t - z'_t a_{t-1} \quad (27)$$

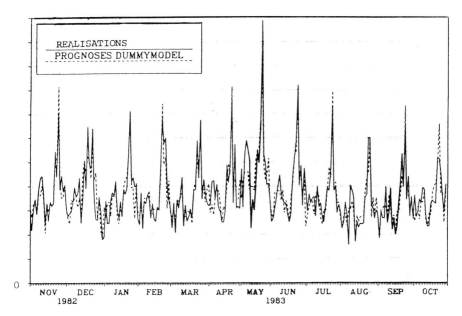

Fig. 4. Daily prognoses of an inflow component.

The criterion is maximization of the likelihood

$$\log L(y) = -\frac{n}{2} \log 2\pi - \frac{n}{2} \log \sigma^2 - \frac{1}{2} \sum_{t=1}^{n} \log f_t - \frac{1}{2\sigma^2} \sum_{t=1}^{n} v_t^2/f_t \quad (28)$$

Where the maximum likelihood estimate of σ^2, the variance of u_t is

$$\hat{\sigma}^2 = \frac{1}{n} \sum_{t=1}^{n} v_t^2/f_t \quad (29)$$

As in the estimation algorithm the variances σ_w^2, σ_m^2 and σ_y^2 are normalized. These should after optimization be multiplied by (29).

Most relations in the algorithm are rather simple because z_t and R_t contain many zero elements. Unfortunately the large matrices $P_{t/t-1}$ and P_t cannot be simplified; due to equation (24) they contain no zero elements. This is a serious handicap when for instance yearly effects are considered; a state vector of 47 is big enough. Things would simplify also when there would be a steady state solution as in most time series models. This occurs when z and R do not depend on t, leading to limiting values of $P_{t/t-1}$, P_t and f_t not depending on t. But z and R depend on t, which is precisely the reason that traditional timeseries techniques run into trouble.

Fortunately our 47 dummies perform well as may be seen from Fig. 4, giving real forecasts of one of the inflows conditional upon monthly aggregates.

8. Comparison of the two models and estimation of possible improvements

Having developed the 'Procrustes' and the Kalman-filter model we spent a rather long time finding out which of them performed best. For real daily forecasts (two years were used) the Kalman-filter won. But our impression was that the need of backtransformation of the Procrustes forecasts made this model lose: around paying days the peaks in the money flows were too strongly smoothed. The Kalman filter model performs better for inflow series with salary payments: the mean absolute forecast error is about 10% lower. For the outflow percentage however the Procrustes model performed (marginally) better.

The most striking difference between the two models is that the Procrustes model uses 262 degrees of freedom (252 dummies, 2 update parameters and 8 fixed dummies) and the Kalman filter only 51 (47 dummies, 1 constant and 3 update parameters). This is due to the fact that yearly effects were not significant. One would expect such differences to have more effect. Perhaps this expectation is based on too much confidence in statistical tests. Recent discussions hardly justify this. Chow (1983, ch. 9) ends up with a revised version of Akaike's Information Criterion, but Leamer (1983, pp. 310 and 316) rather convincingly criticizes the AIC. There seems to be some space left for alternative procedures in selecting the dummies.

To know whether it is worthwhile to work further on the model we developed a simple method to estimate the variance of the forecast error of a model that combines the good features of both models. We use the real forecast errors and suppose

$$V_1 = R + u_1 + \zeta \\ V_2 = R + u_2 + \zeta \tag{30}$$

With u_1, u_2 and ζ normally distributed errors. R is the realization; V_1 and V_2 are the forecasts with the models 1 and 2. ζ is the error that both forecasts have in common. u_1 and u_2 are the specific errors of both models. By the chosen construction all three errors may be supposed to be independent. It has no use to discern in ζ components that may not be forecast at all and components that may not be forecast with models 1 or 2; the outcomes remain the same.

The 'ideal' model has a forecast error σ, and

$$\text{Var}(\zeta) = \text{Cov}(V_1 - R, V_2 - R) \tag{31}$$

For an important inflow series for instance we got

$$\begin{aligned} \sigma_1 &= \text{s.e. of forecasts model } 1 = 0.85 \\ \sigma_2 &= \text{s.e. of forecasts model } 2 = 0.78 \\ \varrho &= \text{corr of forecasts models 1 and } 2 = 0.71. \end{aligned} \tag{32}$$

It follows that the standard error of ζ is 0.684, 12% better than that of model 2. We even derived that with 95% certainty this improvement is between 8% and 16%. Of course it is not yet clear how this improvement must be realized. A well-known alternative is to combine both forecasts:

$$V^* = \lambda V_1 + (1-\lambda)V_2 \tag{33}$$

the variance of this forecast is

$$\lambda^2 \sigma_1^2 + (1-\lambda)^2 \sigma_2^2 + 2\lambda(1-\lambda)\varrho\sigma_1\sigma_2 \tag{34}$$

which has a minimum for

$$\lambda = (\sigma_2^2 - \varrho\sigma_1\sigma_2)/(\sigma_1^2 + \sigma_2^2 - 2\varrho\sigma_1\sigma_2) \tag{35}$$

For the estimates from (31) this gives $\lambda = 0.35$, the variance in (33) becomes .557, the standard error .746, which is only 4% lower than the standard error of model 2. So combining the two models is not very useful. Some further research however might – as always – help.

9. Acknowledgement

This research was done in narrow cooperation with a group of people of the Postal Cheque and Clearing Service (now Postbank). I thank H.C. Dijkman and P.C. Tolk for their useful ideas, H.J. Kooiman for excellent programming and in particular H.W.M. Jansen who did a large part of the research of the Kalman-filter models.

Bibliography

Bell WR and Hillmer SC. 1983. Modelling time series with calendar variation. *Journal of the American Statistical Association*, 78: 526 534.

Chow GC. 1983. *Econometrics*. New York: Mc.Graw-Hill.

Cramer JS. 1969. *Empirical Econometrics*. Amsterdam: North-Holland Publishing Co.

Gersch W and Kitagawa G. 1983. The prediction of time series with trends and seasonalities. *Journal of Business & Economic Statistics*, 1, 253–264.

Harrison PJ and Stevens CF. 1976. A Bayesian approach to short term forecasting. *Journal of the Royal Statistical Society Series B*, 38: 1325–1333.

Harvey AC. 1981. *Time series models*. Oxford: Philip Allan.

Harvey AC and Todd PHJ. 1983. Forecasting economic time series with structural and Box-Jenkins models: a case study. *Journal of Business & Economic Statistics*, 1: 299–315.

Harvey AC. 1984. A unified view of statistical forecasting procedures. *Journal of Forecasting*, 3: 245–275.

Leamer EE. 1983. Model choice and specification analysis, ch. 5 of *Handbook of Econometrics*, 1. Amsterdam: North-Holland Publishing Co.

Pfefferman D and Fisher JM. 1984. Festival and working days prior adjustment in economic time series. *International Statistical Review*, 50: 113–124.

Pierce DA, Grupe MR and Cleveland WP. 1984. Seasonal adjustment of the weekly monetary aggregates: a model-based approach. *Journal of Business & Economic Statistics*, 2: 260–270.

Vos AF de. 1975, 1984. *Seasonal adjustment of unemployment figures; criteria and models.* Dissertation University of Amsterdam (1975, Dutch) and unpublished paper (1976, Free University, Amsterdam).

Vos AF de. 1980. *Univariate time-series analysis,* lecture notes. Amsterdam: Interfaculteit Econometrie, Free University.

Vos AF de. 1983. The specification of relations between time series, specifically money, income and interest (Dutch). *Kwantitatieve Methoden,* 10: 10–32.

Young AH. 1968. Linear approximations to the Census and BLS seasonal adjustment methods. *Journal of the American Statistical Association,* 63: 445–471.

C. Income

Inductive analysis from empirical income distributions

ARNOLD H.Q.M. MERKIES

1. Introduction

According to Lange (1968) the first econometric study has been Pareto's analysis of the personal income distribution. It appeared first in the Giornale degli Economisti of January 1895. Over the last twenty five years the personal income distribution has also been one of the topics of interest of Cramer, see Cramer (1969, 1976, 1978) and Ransom and Cramer (1983). Most authors deal with the problem from a normative point of view. They compare the given distribution with some preconceived idea of how the distribution should look like. Cramer is more at home among positivists who aim at a concise description of actual distributions.

The past twenty five years have also shown increased specialisation in econometrics; more recently in the Netherlands accompanied by a shift from free research to projects evaluated by committees. These judge, among others, the significance of scientific output for the progress of science or society. Cramer was also active in this area, see Cramer (1985).

The question discussed in this paper is what progress has been made in the area of the personal income distribution since Cramer's study of 1969. I will do so from the point of view of the general practitioner with some eye on the actual significance of this research. There are several questions involved, see also Rivlin (1975):
1. How can we describe the personal income distribution.
2. How can we explain its shape.
3. What shape do we prefer.
4. How can we turn the actual shape into the preferred one.
5. What are the effects of changes in the income distribution.

In Section 2 I will clarify the issues with a Pareto distribution. In Section 3 I will concentrate on question 1. in discussing alternatives but with an eye on the remaining questions. Section 4 summarizes.

2. The Paretian income distribution in economic analysis

A concise description of the personal income distribution is provided by some statistical distribution. The standard example is the Pareto distribution, described by the density

$$f(y^*) = \alpha \cdot y^* \cdot \exp\{-(1+\alpha)\} \quad y^* > = 1, \ \alpha > 0 \qquad (2.1)$$

where $y^* = y/y_o$ and y_o is some lower bound below which (2.1) is no longer valid, see Cramer (1969, par. 44). (2.1) is also called the strong Law of Pareto.

The first question is what empirical distribution do we want to approximate by a Pareto distribution. This concerns the definition of the variables. Cramer (1969) indicates in his paragraphs 39 and 40 the relevant distinction between spending units and earning units. Our questions three and five above refer to the former, whereas question two and four actually deal with earning units. Distributions may vary with the concept used. Hence if after finding a satisfactory approximation we want to continue with question two we may prefer a different start than when dealing with question three. The pure empiricist will usually begin with whatever frequency distribution is available and hopes to answer subsequent questions by a process of transformation of variables and discompounding distributions. Alternatively some of these steps may be incorporated in advance e.g. by transforming measured income to a 'more appropriate' income concept, by excluding heterogeneous parts – minimum-wage earners, part-time workers, recent graduates and old-age pensioners – or by disaggregating the distribution e.g. over occupational groups (Harrison, 1981). If prior exclusion is not possible it is also suggested to approximate the empirical distribution by a compound with Pareto as one of the ingredients. The other part may then be ascribed to the unobserved factors such as measurement errors (Ransom and Cramer, 1983) or group variation (Lebergott, 1959). Ransom and Cramer (1983) seems to be the only source which makes use of Pareto in such a compound and they do not find it very suitable. But if we compare empirical distributions with compounds we have in fact already entered the field of alternative distributions to compare with our data. The question of discompounding is more a matter of explanation.

Even if one succeeds in finding a satisfactory approximation there remains the question of stability over time. Tertiary income has become more important (Social and Cultural Planning Bureau of the Netherlands, 1981). Development of community life and emancipatory movements have interfered with the definition of household and its principal earner. Characteristically Dronkers and Bakker's (1981) fourth important factor in explaining the incomes of males from Groningen is 'the occupational level of his (mostly feminine) partner'. It is not clear in advance whether the 'Natural Law' is meant to include these shifting definitions as well.

A subsequent matter is the estimation of the parameters y_o and α. Quandt (1966) found no serious differences in measuring the parameters of the Pareto distribution by least squares, moments, quantiles or maximum likelihood, but Aigner and Goldberger (1970) proposed to estimate α and y_o from the regression

$$f_i = p_i(\alpha, y_o) + \varepsilon \qquad (2.2)$$

where f_i is the empirical frequency in income bracket $[y_{i-1}, y_i)$, p_i the theoretical frequency from Pareto and ε the error of approximation which incorporates all kinds of other factors that have affected the particular frequency f_i and may be normally distributed. Nowadays the parameters are usually estimated by maximizing the multinomial likelihood function

$$L(\alpha, y_o) = N!\, \Pi_i (p_i(\alpha, y_o))^{n_i}/n_i! \qquad (2.3)$$

see McDonald (1984) with the implicit assumption that the difference between the empirical and the theoretical distribution is completely caused by sampling variation. A crucial question is also what criterion is used for evaluating the approximation. Using a minimum – chi – square method as performed by Chesher (1979), Kloek and van Dijk (1978), McDonald and Ransom (1979) and Harrison (1981) invariably leads to the rejection of whatever null hypothesis is used. Rather than the suggestion of Kendall and Stuart (1969) to reduce the power of the test in such cases or the introduction of a new test for the goodness of fit as proposed by Gastwirth and Smith (1972) one may well accept that the measured distribution as such is not as assumed. The difficulty may arise from ignoring all other factors that are contained in ε. This suggests compounds e.g. of Pareto with a distribution attached to ε such as attempted by Ransom and Cramer (1983). We have already seen that Pareto is not successful in this respect. But even with superior compounds we may not expect a complete explanation. All comparisons are relative to each other. Of course we are now in the same unfavourable position as in the usual regression analysis where no proper criterion exists to choose between p_i and an alternative probability π_i, but this is a matter which ultimately belongs to the basic philosophy of our science, see Leamer (1978).

We may conclude that starting an analysis of the income distribution by looking for an appropriate descriptive statistical distribution may be a serious scientific endeavour, but the empirical evidence makes it questionable whether the Pareto distribution is the most proper choice. Its main defective is that 'it decreases monotonically to the right, and it therefore applies to a part of the income distribution well beyond the modal income only' (Cramer, 1976), see also Kakwani (1980). According to Pen and Tinbergen (1976) the Pareto distribution only covers the upper 35% of the distribution. This is the 'conventional wisdom' that the upper tail of the empirical distribution may be described by (2.1). Alternatives are discussed in Section 3.

Our second question is what process generates the Pareto distribution and how can this process be explained in economic terms. One possibility is to do so in the line of Champernowne (1953) and Lydall (1968). However, in spite of various attempts during the past seventeen years it remains 'an unsolved problem' – Cramer (1969). Since 1969 Champernowne's Markov model completed with additional estimation techniques has been proved useful in generating actual income distributions, see e.g. Hartog (1973), but no proof has been given that the estimated transition probabilities are capable of generating a Pareto distribution. Another possibility is to try to discompound the Pareto distribution in constituting distributions that are easier to explain. We have seen that Pareto is a bad compounder; even a generalized Pareto distribution is considered a poor start in this respect, see e.g. Creedy (1977). Finally one may aim only at an explanation of the 'conventional wisdom'. Mandelbrot (1960) offered a mathematical explanation of a weak (asymptotic) version of Pareto's Law

$$\lim_{y \to \infty} \frac{1 - F(y^*)}{y^*} = 1 \qquad (2.4)$$

Question three raised above concerns the normative view. This is often cast in terms of Lorenz curves or inequality measures. The Lorenz curve corresponding with the Pareto distribution is

$$L(p, \alpha) = 1 - (1 - p)^{(\alpha-1)/\alpha}. \qquad (2.5)$$

It is clear that two such Lorenz curves for different alpha's do not cross. Hence, if incomes are indeed Pareto distributed, actual income distributions could always be completely ordered. This is equivalent with the following propositions, see Rothschild and Stiglitz (1973):

If Pareto is the true distribution and more equality is preferred over less then

a) the Social welfare function is a quasi-concave symmetrical function of individual incomes

b) the only requirement that redistributions must fulfil to increase welfare is the Dalton-Pigou condition, i.e. the only requirement for redistributions is that the payer is richer than the receiver.

Many inequality measures would become superfluous as they are then simple transformations of Pareto's α e.g. Gini's constant would be $\alpha/(\alpha - 1)$, the Gini-index $1/(2\alpha - 1)$ and Theil's inequality-coefficient $T = 1/(\alpha - 1) - \ln\{\alpha/(\alpha - 1)\}$. Normalists do not hesitate, however, to discuss a large variety of other inequality measures see e.g. Tinbergen and Pen (1976), Sen (1972) or Bartels (1977). This is an implicit rejection of the Pareto distribution. From this point of view one wonders why normalists have not paid more attention to the question in what respect the Pareto distribution is inadequate.

Question four has induced various authors e.g. Kakwani (1980), Korzec (1981), de Wolff (1976) to turn away from the stochastic theory with the allegation that it does not give the policymaker any instrument for change. This is only partly true. Of course individual items are not controllable in a stochastic theory, but the distribution and its parameters are. We will show in section three that what these authors actually do is concentrating on an explanation of (one of) the parameters of some distribution by explanatory factors. This leaves the random process as such unexplained but as soon as a satisfactory explanation of the generation of the random process is available, one may also consider to interfere with this generation process. If Pareto is the proper distribution we are interested in how the parameters y_o and α depend upon control variables such as taxes. I do not know of any studies in this area.

Induced changes in the income distribution may have effect on a large number of other variables. I concentrate on the effects of income redistributions on consumer expenditure. This requires – Cramer (1976, p 126) – the evaluation of

$$y = \int_0^\infty g(y) \cdot f(y|\alpha, \beta, \ldots) \, dy \qquad (2.6)$$

where y is income per head, $g(y)$ is the individual Engel curve and $f(y|\alpha, \beta, \ldots)$ the income distribution with parameters $\alpha, \beta, \ldots$ etc. A quadratic Engel curve would thus connect income per head with the mean and the second moment of the Pareto distribution

$$y = a + b\, y_o\, \alpha/(\alpha - 1) + c y_o^2\, \alpha/(\alpha - 2) \qquad (2.7)$$

and income redistribution measures can be evaluated by

$$dy/d\alpha = -b\, y_o/(\alpha - 1)^2 - 2c\, y_o^2/(\alpha - 2)^2. \qquad (2.8)$$

With α not far from 2 this effect will mainly depend upon the value of c. For linear Engel curves (c = 0) redistribution measures ($d\alpha < 0$) will have a positive effect on demand of non-inferior goods. For the usual sigmoid Engel curve the quadratic is only an approximation which gives $dy/d\alpha < 0$ for $c > 0$ i.e. for goods in their take-off period (luxuries?) and $dy/d\alpha > 0$ for $c < 0$ i.e. for goods more near their point of saturation. More elaborate answers to this question would require different Engel curves for different income brackets which means variation of b over incomes. Unfortunately such reflections with a Pareto distribution remain academic in view of the partial coverage of the income distribution by Pareto.

From the considerations above I conclude that a Pareto distribution is a poor tool in explaining income distributions:
a) It covers only the top 35%.
b) It has no proper mode.

c) Its generation process is hard to explain in economic terms.
d) It is an awkward compounder.
e) It is too restrictive for normalists.

Of course the conclusions about the empirical results with a Pareto distribution remain true: 'Ces résultats sont très remarquables. Il est absolument impossible d'admettre qu'ils sont dus au hasard. Il y a bien une cause qui produit la tendance des revenues a se disposer suivant une certaine courbe' (Pareto, 1896 p 377). As the coincidence refers only to the top of the distribution we may replace (2.1) by the weak version of Pareto's Law or by the weak-weak version proposed by Esteban (1986):

$$\lim_{y \to \infty} \overrightarrow{\pi(y)} = -\alpha \tag{2.9}$$

where $\pi(y)$ is the income share elasticity at income y. In passing to alternative distributions we may ask whether these also satisfy (2.9). Esteban even interprets (2.9) as a theoretical requirement of the preferred distribution, which is testable. In my view this is in conflict with the empirical base of (2.9).

In view of the limited significance of the Pareto distribution for applications the explanation of its having a proper tail will probably not evoke much enthusiasm (and financial support) for intensified research.

3. Alternatives

Without Pareto we have to tackle our five questions with alternative distributions. A large number of these is available, see e.g. Maddala and Singh (1977). McDonald (1984) shows that many of these can be derived from either generalization of the beta distribution he presents. Dagum (1977) classifies distributions in three categories:
a) Functions that arise from a stochastic process.
b) Functions that are merely descriptive.
c) Functions that arise as the solution of a differential equation.

The third approach is very useful in classifying and generalizing distributions, see also Esteban (1978). We are mainly interested in group a) but promotion of distributions from b) to a) can always happen. I will not cover all alternatives neither all aspects required to meet with our five questions but only discuss some aspects of some distributions.

The lognormal distribution is introduced as a description of the income distribution by Gibrat in 1931 and independently by van der Wijk in 1939. The main reference is Aitchison and Brown (1969). The lognormal is a special case of the generalized gamma density, which in its turn can be derived from either of McDonald's beta functions. It pretends to approximate the complete distribution rather than some part of it and indeed fairly successfully apart from its too skinny right-hand tail. In old-fashioned words the empirical distribution is leptocurtic as compared to the lognormal distribution. The skinny tail is responsible for the failure of the lognormal to satisfy the weak-weak Law of Pareto. Its income-share elasticity equals $1 - a/b + 1/b \ln y$, which clearly does not converge to a constant value if y approaches infinity. Its role in normative theories is comparable with Pareto's, the lognormal parameter σ taking the place of Pareto's α. The foundation of the distribution upon Central Limit Theorems is not unquestioned, see Sahota (1978). No conclusive explanation of its constant variance is presented. The blending capacity of the lognormal is not impressive either but for the convolution of two lognormal distributions. The latter capacity is used by Cramer (1969) to derive aggregate demand of attributes that have non-differentiable demand curves if considered individually.

The point raised in Section 2 on the role of stochastic theory specially refers to the lognormal distribution. Assume income is lognormally distributed:

$$y \simeq \wedge(\mu, \sigma) \tag{3.1}$$

or

$$\ln y = \mu + \varepsilon \text{ with } \varepsilon \simeq N(0, \sigma). \tag{3.2}$$

Explanation of μ by years of schooling gives

$$\ln y = \vartheta + \gamma S + \varepsilon \tag{3.3}$$

which is standard human capital theory.

It is clear that stochastic theory and human capital theory are not competitors but allies. Without an underlying stochastic theory no explanation is available for the choice of ln y in the left-hand side of (3.3) nor for the assumption of a normal distribution of ε. In other words human capital theory has still to verify whether the income distribution – conditional upon schooling – is lognormal. Alternatively assume that for each education level the income distribution can be described by $f(y|\varphi)$ where φ is a vector of parameters one of which, φ_1 say, is connected with schooling and let the distribution of schooling be given in terms of the parameter φ_1 by $g(\varphi_1|\alpha, \beta, \ldots)$ then the (marginal) income distribution for the population is

$$h(y|\varphi_2, \ldots, \varphi_n, \alpha, \beta, \ldots) = \int f(y|\varphi) \cdot g(\varphi_1|\alpha, \beta, \ldots) \, d\varphi_1 \tag{3.4}$$

Distribution $h(y| \ )$ depends upon the parameters of schooling $\alpha, \beta, \ldots$ one of which presumably will be connected with the number of years of schooling. The inductive approach demands whether the function $h(y| \ )$ is lognormal and subsequently whether any of its parameters depends upon schooling or even whether a schooling function $g(\varphi_1| \ )$ can be disentangled from decomposing h(y|). We conclude that human capital theory is more deductive, specifies (3.3) before (3.1), takes schooling as non-stochastic and deals with distributions conditional upon variables which can be observed in principle.

The stochastic approach starts with (3.1), is concerned with conditional distributions upon parameters or with marginal distributions if a prior distribution of the parameters is assumed. Conditionality upon the parameters of priors of course remains.

A final remark is relevant with respect to the equivalent of (2.7) for the lognormal distribution as the latter was used by Cramer in this context. Substitution of the first and second moment of the lognormal distribution gives

$$y = a + b \exp\{\mu + 0.5\sigma^2\} + c \exp\{2\mu + 2\sigma^2\} \tag{3.5}$$

$$dy/d\sigma = b\sigma \exp\{\mu + 0.5\sigma^2\} + 4c \exp\{2\mu + 2\sigma^2\} \tag{3.6}$$

giving Cramer's redistribution elasticity of demand as

$$s = -1/y \, dy/d\sigma = \sigma[\{a + 3c \exp(2\mu + 2\sigma^2)\}/y - 1]. \tag{3.7}$$

For $c = 0$ the expression is even more simple but linear Engel curves may be too restrictive. Cramer used loglinear and semilog functions. Gorman (1981) showed that standard demand theory excludes more than three independent income terms in the Engel curve. He also claims to have derived an exhaustive list of possible terms, but his list should be used with care, for a logquadratic function as used by Deaton (1981) may be questioned, see Merkies and van Daal (1985). Hence, although (3.7) and Cramer's specifications are not limitative the number of similar results does not seem to be very large.

Other contenders in the race are the gamma distribution and the distribution put forward by Singh and Maddala (1976). The latter was criticized by Cramer (1978) on its interpretation and its descriptive power. Cramer's statement that the number of parameters of the SM distribution is three, whereas the gamma density has only two is clearly shown in McDonald's (1984) relation tree. Peculiarly enough McDonald concludes 'The Sing-Maddala (or Burr) distribution provided a better fit than the generalized beta of the first kind (four parameters) and all of the two and three parameter models considered. The ... Function ... greatly facilitates estimation and analysis of results'. On the other hand the *generalized* gamma is according to Esteban (1986) the only function, which satisfies her three hypotheses: weak-weak Pareto, interior mode and constant declining income-share elasticity.

In view of Esteban's statement I will devote some words on results with the simple gamma density, anticipating future discussions. The generation process of this distribution can hardly be seen as the solution in explaining the empirical distribution. The usefulness of the gamma distribution lies in its capacity to be decomposed. Assume for instance that the income distribution for some homogenous group can be described by a gamma function

$$f(y|\mu, \gamma, \vartheta) = \frac{\mu^{\gamma+\vartheta+1} y^{\gamma+\vartheta} e^{-\mu y}}{(\gamma+\vartheta)!}, \quad y \geq 0, \mu > 0 \tag{3.8}$$

with γ and ϑ nonnegative integers so that $E(y) = (\gamma + \vartheta + 1)/\mu$.

Take first $\vartheta = 1$ and let γ be connected with the occupation (or education level or something similar) and assume that γ follows a Pascal process with

$$g(\gamma|\delta) = (\gamma + 1)(1 - \delta)^2 \delta^\gamma \quad \gamma = 0, 1, 2, \ldots, 0 < \delta < 1 \tag{3.9}$$

then evaluation of (3.4) gives

$$h(y|\delta, \mu) = \sum f(y|\mu, \gamma) g(\gamma|\delta) = [\mu(1-\delta)]^2 \, y \, e^{-(1-\delta)\mu y} \tag{3.10}$$

which means that also the aggregate income distribution is gamma. Hence if we start with (3.10) we may consider decomposition into (3.9) and (3.8).

Another possibility is to start with (3.8) for general ϑ and replace (3.9) by a Poisson process

$$g(\gamma|\delta) = \frac{e^{-\delta} \delta^\gamma}{\gamma!}. \tag{3.11}$$

The result is

$$h(y|\delta, \mu, \vartheta) = \mu \, e^{-\mu y - \delta} (\mu y/\delta)^{\vartheta/2} A_\vartheta(2\sqrt{\mu y \delta}) \tag{3.12}$$

with $A_k(2\sqrt{\mu y \delta}) = \sum_n 1/n! \, (n+\vartheta)! \, (\sqrt{\mu y \delta})^{2n+\vartheta} \quad \vartheta = 0, 1, \ldots$

the 'modified' Bessel function.

The latter shows that it is easier to construct an aggregate from the gamma density than to follow the reverse process. The empirical econometrician who would like to induce laws from data would rarely start from a comparison of his empirical data with (3.12). The theory lends itself more to the study of the effects of e.g. the occurrence of occupational groups on the income distribution. This affects the attractiveness of the gamma distribution for those who like to start from pure induction. Moreover the gamma distribution for the individual groups (3.8) is still to be explained.

Harrison's (1981) study of such disaggregated distributions did not give attention to the gamma distribution. I doubt whether his conclusions would have been different. Lebergott (1959) is of the opinion that decomposition of the actual distribution over occupational groups would have resulted into $f(y|\ldots)$ to be a normal distribution. Unfortunately the normal distribution like its logarithmic transformation does not possess such nice compounding properties. This means that such allegations can only be tested by functions $h(y|\ldots)$ that arise from numerical integration, see e.g. Ransom and Cramer (1983) and van Praag, Hagenaars and van Eck (1983). If the compound can only be found by numerical integration the possibility of starting from the aggregate and disaggregating to find the composing parts is excluded. Our conclusion is that so far the mixing properties of the gamma distribution have not been of much help in inductive settings.

I will end this by referring again to the study of the effects of income redistribution on consumer demand. Cramer (1976) used the gamma distribution with semilog and loglinear individual Engel curves to derive an Engel curve for the aggregate. The alternative with a quadratic Engel curve results if $f(y|\alpha, \beta, \ldots)$ in (2.6) is substituted by (3.8). With $\vartheta = 0$ it gives

$$y = a + b\,y + c(\gamma + 2)/(\gamma + 1)\,y^2 \qquad (3.13)$$

see van Daal and Merkies (1984 p 161).

The lengthy discussions above have not given any verdict on whether the approximation of the empirical distribution by some statistical distribution is a useful start to answer questions around the income distribution. The frequent use of the method of moments in estimating the parameters of such distributions already suggests an alternative approach. Replacing the density function by its moment-generating function is equivalent and for all practical purposes we can restrict ourselves to only a limited number of moments. The problem of the effects of income redistribution on consumer expenditures needs only a small number of moments in view of the results of Gorman (1981). In fact we do not need to continue beyond the second moment of the income distribution, although prior transformation · of to ln y, y ln y or $(y^k - 1)/k$ (the Box-Cox transformation) remain possible alternatives. For equal individual Engel curves we can thus start from

$$y = a + b\,y + c\,y^2. \qquad (3.14)$$

Comparison with (2.7) and (3.13) reveals that in these previous formulae y or c are replaced by a parameter expression. Another approach is to do a regression of type (3.14) and derive results on income redistribution from the parameter c. From van Daal and Merkies (1984 p 243) we obtain Table 1.

From this we may conclude that there has been a negative influence of income redistribution on consumer demand as the time pattern of the coefficient of variation (which was used instead of the second moment) showed a general decline from 1.76 in 1948 to 1.50 in 1975.

Table 1. Estimates of c for the Netherlands 1948–1975.

Food, beverage and tobacco	.371	(.02)
Durables	.396	(.04)
Other services	.233	(.01)

As remarked before (3.14) is not the only specification left to us by Gorman but I doubt whether transformations would affect the conclusions considerably.

As another judgement on the use of moments instead of distributions I present the first five moments of the income distributions of the United States from 1954 to 1970. The data are compiled by van Renswoude (1985) from the 'The Statistics of Income, individual income tax returns, U.S. Treasury Department, Internal Revenue Service' in the years 1959, 1965 and 1970 and the 'Statistical Abstract of the United States, U.S. Department of Commerce, Bureau of the Census'. The data refer to after-tax income. The moments are scaled by taking square root, cubic root, etc.

Study of these moments is equivalent to the study of distributions. For instance if a gamma distribution would be the proper description the first five moments are

$$\beta^k (\gamma + 1)(\gamma + 2) \ldots (\gamma + k) \qquad k = 1, 2, \ldots, 5 \qquad (3.15)$$

with $\beta = 1/\mu$. Recalculating them to central moments, see Kendall and Stuart (1969), and equating them to the values above gives estimates of γ and β and an indirect evaluation of the gamma distribution. I have not gone into that, but to obtain a first view on this issue I have calculated the correlations between the central moments.

Table 2. Central moments, Income Distribution of the United States (thousands of dollars).

Year	Moments 1	2	3	4	5
1954	3.62	3.97	14.83	39.57	73.25
1955	3.80	4.37	17.13	45.13	82.56
1956	4.01	4.46	17.12	45.38	83.12
1957	4.16	4.39	15.34	40.31	73.95
1958	4.22	4.61	18.02	48.76	90.27
1959	4.48	5.04	19.83	52.67	96.36
1960	4.57	5.00	18.97	49.91	90.93
1961	4.73	5.43	21.22	54.80	98.56
1962	4.90	5.36	20.17	52.75	95.65
1963	5.07	5.55	21.01	54.97	99.58
1964	5.41	6.16	23.59	60.68	108.71
1965	5.67	6.83	27.01	68.56	121.56
1966	5.94	6.93	25.95	65.71	116.63
1967	6.22	7.36	26.24	64.54	112.86
1968	6.54	7.94	29.03	70.34	121.45
1969	6.89	8.06	29.20	70.62	121.59
1970	7.45	7.72	26.10	66.24	118.00

Table 3. Correlations between the central moments of Table 2.

	1	2	3	4	5
1	1	.977	.930	.924	.916
2		1	.982	.969	.956
3			1	.995	.986
4				1	.998
5					1

Although we have seen that in general no three parameter approximation other than the Singh-Maddala distribution is satisfactory, see also Kloek and van Dijk (1976), the correlations of Table 3 are high enough to retain only a small number of parameters if time effects of the income distribution on consumer outlay are studied. Moreover expansion to a higher number of moments meets with considerable difficulties, see Kloek and van Dijk (1977). We have already seen that within the classical study of consumer demand these are not needed either. The question remains what factors are responsible for the development of these moments, especially what causes underly the increase of the standardized central moments with the increase of the degree.

4. Summary

In the previous Sections I have tried to discuss prospects of finding proper descriptions of empirical income distributions. It is not to be expected that 'the' proper description will be discovered. Including more complex functions and more parameters will give better descriptions but not always better insight in underlying forces. Therefore it remains wise to select a favourite distribution with an immediate eye on such matters as tail behaviour, mode, behaviour of income-share elasticity, (dis)compounding abilities, generation process etc. A large number of parameters is in itself no guarantee for obtaining more information. For the pure inductive approach it remains sensible to work with small parameter approximations for actual distributions. One should only abstain from such criteria as χ^2 that pretend that a complete description can be given and that actual differences are only due to sampling errors. For studying effects of changes in the income distributions an analysis in terms of (empirical) moments may suffice. The final answer to the question whether the inductive approach is promising enough is left to the readers. I hope that there will be many of them.

Bibliography

Aigner DJ and Goldberger AS. 1970. Estimation of Pareto's law from grouped observations. *Journal of the American Statistical Association*, 65: 712–723.

Aitchison J and Brown JAC. 1969. *The lognormal distribution*. With special reference to its uses in economics, Cambridge: Cambridge University Press.
Atkinson AB. 1970. On the measurement of inequality. *Journal of Economic Theory.* 2: 244–263.
Atkinson AB. 1973. *Wealth income and inequality*. Harmondsworth: Penguin Education.
Bartels CPA. 1977. *Economic Aspects of Regional Welfare. Income distribution and unemployment.* Leiden: Martinus Nijhoff.
Champernowne DG. 1953. A model of income distribution. *Economic Journal,* 68: 318–351.
Chesher A. 1979. An analysis of the distribution of wealth in Ireland. *Economic and Social Review,* 11: 1–17.
Cramer JS. 1969. *Empirical Econometrics*. Amsterdam: North-Holland Publishing Co.
Cramer JS. 1976. The effect of income redistribution on consumer demand, in: *Relevance and precision. From quantitative analysis to economic policy. Essays in honour of Pieter de Wolff,* edited by Cramer JS, Heertje A and Venekamp P. Alphen aan de Rijn/Amsterdam: Samson/North-Holland Publishing Co.
Cramer JS. 1978. A function for size distribution of incomes: comment, note. *Econometrica,* 46: 459–460.
Cramer JS. 1985. *Een profiel van het academisch onderzoek in Nederland in 1983.* Note AE NI/85(R) Faculty of Actuarial Science and Econometrics, University of Amsterdam.
Creedy J. 1977. Pareto and the distribution of income, note. *Review of Income and Wealth,* 23: 405–411.
Cronin DC. 1979. A function for size distribution of incomes: a further comment, note. *Econometrica,* 47: 773–774.
Daal J van and Merkies AHQM. 1984. *Aggregation in economic research.* Dordrecht: D. Reidel Publishing Co.
Dagum C. 1977. A new model of personal income distribution: specification and estimation. *Economie Appliquee,* 30: 413–437.
Deaton AS. 1981. *Essays in the theory and measurement of consumer behaviour.* Cambridge: Cambridge University Press.
Dronkers J and Bakker B. 1981. Leren maakt het leven rijker, ouderlijk milieu, genoten onderwijs, de verdere levensloop en de hoogte van het netto gezinsinkomen. *Intermediair,* 17, 57–65.
Esteban JM. 1978. A general density function for the distribution of income. Report Facultad de Ciencas Economicas y Empresariales, Universitad Autonoma de Barcelona.
Esteban JM. 1986. Income-share elasticity and the size distribution of income. *International Economic Review,* 27: 439–444.
Gastwirth JL and Smith JT. 1972. A new goodness-of-fit test. *Proceedings of the American Statistical Association,* Business and Economic Statistics session, 320–322.
Gorman WM. 1981. Some Engel curves, chapter 1 in: *Essays in the theory and measurement of consumer behaviour,* edited by Deaton A. Cambridge: Cambridge University Press.
Harrison A. 1981. Earnings by size: A tale of two distributions. *Review of Economic Studies,* 48: 621–631.
Hartog J. 1973. Een vergelijking van inkomensmobiliteit naar beroepsgroepen. *Praeadvies voor de Vereniging voor de Staathuishoudkunde,* blz. 33–62.
Kakwani NC. 1980. *Income inequality and poverty. Methods of estimation and policy applications.* A World Bank publication. New York: Oxford University Press.
Kendall MG and Stuart A. 1969. *The advanced theory of statistics.* 3rd edition, London: Griffin.
Kloek T. and Dijk HK van. 1976. Efficient estimation of income distribution parameters. Report 7616/E Econometric Institute, Erasmus University Rotterdam.
Kloek T and Dijk HK van. 1977. Further results on efficient estimation of income distribution parameters. *Economie Appliquee,* 30: 439–459.
Kloek T and Dijk HK van. 1978. Efficient estimation of income distribution parameters. *Journal of Econometrics,* 8: 61–74.

Korzec M. 1981. Inkomensverdeling: Ideologieen, begrippen en theorieen. *Intermediair*, 17: 31–39.
Lange O. 1969. *Introduction to Econometrics*. Oxford: Pergamon Press.
Lebergott S. 1959. The shape of the income distribution. *American Economic Review*, 49: 328–347.
Leamer EE. 1978. *Specification searches*. Ad hoc inference with nonexperimental data. New York: John Wiley and Sons.
Lydall H. 1968. *The structure of earnings*. Oxford: Clarendon Press.
Maddala GS and Singh SK. 1977. Estimation problems in size distributions of incomes. *Economie Appliquee*, 30: 461–480.
Mandelbrot B. 1960. The Pareto-Lévy law and the distribution of income. *International Economic Review*, 1: 647–663.
McDonald JB and Ransom MR. 1979. Functional forms, estimation techniques and the distribution of income. *Econometrica*, 47: 1513–1525.
McDonald, JB. 1984. Some generalized functions for the size distribution of income. *Econometrica*, 52: 647–663.
Merkies AHQM and Daal J van. 1985. Nataf's theorem, Taylor's expansion and homogeneity in consumer demand. Onderzoeksverslag no. 148 Interfaculteit der Actuariele Wetenschappen en Econometrie, Vrije Universiteit Amsterdam.
Mustert GR. 1974. The development of the income distribution in the Netherlands after the second world war: a Markovian approach. Research Memorandum no. 47 Tilburg Institute of Economics, Department of Econometrics.
Parcto V. 1896. *La courbe de la repartition de la richesse*. Lausanne: Viret-Genton.
Pen J and Tinbergen J. 1976. Hoeveel bedroeg de inkomensegalisatie sinds 1938. *Economisch Statistische Berichten*, 61: 880–884.
Praag BMS, van Hagenaars A and Eck W van. 1983. The influence of classification and observation errors on the measurement of income inequality. *Econometrica*, 51: 1093–1108.
Quandt RE. 1966. Old and new methods of estimation and the Pareto distribution. *Metrika* 55–82.
Ransom MR and Cramer JS. 1983. Income distributions with disturbances. *European Economic Review*, 22: 363–372.
Renswoude GJ van. 1985. Twee consistent geaggregeerde vraagmodellen. Doctoraalscriptie Erasmus Universiteit Rotterdam.
Rivlin AM. 1975. Income distribution – can economists help? Richard T Ely lecture, *The American Economic Review*, 65, 1–15.
Rothschild M and Stiglitz JE. 1973. Some further results in the measurement of inequality. *Journal of Economic Theory*, 6: 188–204.
Sahota GS. 1978. Theories of personal income distribution: a survey. *Journal of Economic Literature*, 16: 1–55.
Salem ABZ and Mount TD. 1974. A convenient descriptive model of income distribution: the gamma density. *Econometrica*, 42: 1115–1127.
Sar NL van der. 1985. Het verband tussen de coefficient van Theil en de variatie-coefficient in samenhang met sociale welvaart. Rapport no. 85.04 Center for Research in Public Economics, University of Leiden.
Sen A. 1973. *On economic inequality*. Oxford: Clarendon Press.
Singh SK and Maddala GS 1975. A stochastic process for income distribution and tests of income distribution functions. *ASA Proceedings of Business and Economic Statistics*.
Singh SK and Maddala GS. 1976. A function for size distribution of incomes. *Econometrica*, 44: 963–970.
Sociaal and Cultureel Planbureau. 1981. *Profijt van de overheid in 1977*. Sociale en Culturele Studies. 1. Staatsuitgeverij 's-Gravenhage.
Wolff, P. de (1976), Theorieen der Persoonlijke Inkomensverdeling. Afscheidscollege Universiteit van Amsterdam, 30 oktober 1976.

Economic growth and the size distribution of income; a longitudinal analysis

MICHAEL R. RANSOM

1. Introduction

Policy makers and economists have long been interested in the effect of economic growth on the size distribution of incomes. Some economists believe that economic development in countries can come only at the expense of greater distributional inequality, suggesting that policies that encourage economic growth should be accompanied by redistributional policies. Others feel that economic growth will naturally result in a more equitable distribution of income, as all groups benefit from growth.

While reliable data on income distribution are scarce, a considerable number of empirical studies have tried to analyze the relationship between economic development and the size distribution of incomes. The findings can be roughly summarized in the following way:

(1) Income is more equally distributed in developed countries than it is in developing countries. This is established by Kuznets (1963), Ahluwalia (1980), Oshima (1962), and others.
(2) In the developed, western countries, income is more equally distributed now than it was in the early part of this century and before. Kuznets (1963) reports this, and Soltow's (1968) study of the British upper class further supports it.
(3) Economic growth in less developed countries is often accompanied by an increase in the concentration of income. For example, Weiskoff (1970) finds this in his study of Puerto Rico, Argentina and Mexico. Berry (1974) reports similar results for Columbia.

The 'model' used to explain these stylized facts, sometimes referred to as the 'Kuznets curve', states that in early stages of development, modernization and economic growth will cause increased concentration of incomes due to increased importance of a relatively few who own capital or who have scarce skills. However, beyond some critical point in the path of development, the distribution of income begins to equalize. Unfortunately, the empirical evidence is less convincing than it could be because of the poor quality of the data on income distribution.

Researchers have acknowledged that much of the data on income distribution is unreliable, and have stressed the tentative nature of their conclusions. In cross-sectional analysis this is particularly problematic, since different countries use different concepts of income, and base their data on different recipient concepts (whether individuals, households, families, or workers). The problem is further aggravated by the fact that the quality of the data varies from country to country.

Yet even with reliable data, there would be conceptual problems with using cross-sectional data to discover the effect of economic growth on the size distribution of incomes, since the economic histories of developed countries differ in many dimensions from those of the less developed countries. For example, many undeveloped countries served as colonies for long periods of their histories. Also, developing countries may have more social, ethnic, linguistic and even geographic barriers to equalizing the income distribution. Cross-sectional analysis tends to attribute all of these differences to the different levels of development.

Analyzing changes in the income distribution over time within individual countries provides a more convincing argument. There have been several careful studies of countries based on longitudinal data. For example, Weiskoff (1970) examined the changes in the income distribution in Mexico, Argentina and Puerto Rico during the growth of the 1950's and 1960's. Berry's (1974) study of Columbia covers a longer span of time. Yet even these studies suffer from the lack of data that are strictly comparable over time.

However, reliable panels of data on income distribution do exist, primarily in the more developed countries. For example, the United States has carried out annual surveys on income distribution since the late 1940's with few changes in concept over that period of time. Canada and Australia also report results of periodic surveys, and consistent longitudinal data are available from several other countries for recent decades. In this paper, I exploit these high-quality longitudinal data to examine the effect of economic growth on the size distribution of income. Emphasizing the changes in individual countries over time mitigates the importance of the differences that underly the concepts and measurement of the data.

2. Data

Data comparability is always a problem in studies of income distribution across countries. There are several dimensions to the problem. 1. The recipient unit: Most studies that are interested in making some kind of welfare comparison have chosen income distributions based on families or households, since in general these are the units where decisions about consumption are made. However, data are often collected for individual workers, or simply for individuals. In this study I have used data on households, where that is possible. 2. The statistical source of the data: Most data are either based on a sample survey of the country or are based on income tax records. Tax data are not strictly comparable with survey data because some groups may be excluded from paying taxes, some income is excluded from tax information, and there is a well defined incentive to underreport income for tax purposes. On the other hand, survey results depend on the wording of questions and concepts used. 3. A third problem comes from different concepts and sources of income. Data on earnings are generally reliable, but information on income from other sources is less accurate. Data sources differ in their treatment of the different sources of income. This is particularly important in studies of income distribution, since some income groups may depend more heavily on one source of income than other groups.

In this study, I analyze changes in the distribution of income in seven developed or moderately developed countries: Australia, Canada, Israel, Taiwan, United Kingdom, United States and West Germany. The sources of the data for the analysis are given in Table 1. This table also points out the differences in the concept of income and recipient that underly the data from the various countries. These differences make comparisons of inequality between countries inappropriate. For example, the UK data are based on tax records which makes it difficult to compare the level of inequality in Britain with that of the United States, which reports survey data. The analysis assumes that the data will reflect changes in the income distribution over time within a country, even though inter-country comparisons are not possible.

To explain the changes in the income distribution, I choose a small set of variables that have been stressed in many previous studies of this type:
1. the growth rate of per capita real Gross National Product (GROWTH).
2. the fraction of labor force in agriculture (AGRI).
3. the fraction of the population under 14 (AGE1).
4. the fraction of the population over age 65 (AGE2).
5. the fraction of gross national product spent by the federal government (GOV).

GROWTH is included for obvious reasons. Some theories point to the differences between the high-wage industrial sector and the low-wage agricultural sector as important sources of income inequality, though this is not likely to be very important for the sample I examine in this paper. AGRI will capture this difference. AGE1 and AGE2 are included in an attempt to capture life-cycle

Table 1. Sources of data.

Australia
   Years included: *1969, 1979, 1982.*
   Sources: *Official Year Book of Australia,* Australian Bureau of Statistics, Canberra, various years.
      Income distribution based on survey data. Recipients are families.

Canada
   Years included: 1951, 1957, 1961, 1967, 1971, 1975.
   Sources: *Canada Year Book,* Minister of Industry, Trade and Commerce. Various years.
      Income distribution based on survey data. Recipients are families.

Israel
   Years included: 1965, 1969, 1974, 1980.
   Sources: *Statistical Abstract of Israel,* Central Bureau of Statistics, various years.
      Income distribution based on survey data. Recipients are households.

Taiwan
   Years included: 1964, 1968, 1972, 1976, 1980.
   Sources: *Taiwan Statistical Data Book,* Republic of China Council for Economic Planning and Development, various years.
      Income distribution based on survey data. Recipients are households.

United Kingdom
   Years included: 1959, 1964, 1967, 1972.
   Sources: Income distribution data from V. George and R. Lawson (eds.), *Poverty and Inequality in Common Market Countries,* London: Routledge and Kegan Paul, 1980. All other data from *Annual Abstract of Statistics,* Central Statistical Office, London, various years.
      Income distribution based on income tax data. Recipient concept is income tax filer, which is typically households.

United States
   Years included: 1947, 1952, 1957, 1962, 1967, 1972, 1977, 1982.
   Sources: *Statistical Abstract of the United States,* U.S. Department of Commerce, Bureau of the Census, various years.
      Income distribution based on survey data. Recipients are families and unrelated individuals (households).

West Germany
   Years included: 1950, 1960, 1968, 1973.
   Sources: Income distribution data are from V. George and R. Lawson (eds.), *Poverty and Inequality in Common Market Countries,* London: Routledge and Kegan Paul, 1980. All other data are from *Statistisches Jahrbuch für die Bundesrepublik Deutschland,* Statistisches Bundesamt, Wiesbaden, various years.
      Income distribution is based on estimates by the Deutsches Institut für Wirtschaftsforschung. Recipient concept is households.

---

differences in the distribution of income. AGE1 also is a good proxy for the population growth rate since high population growth leads to a relatively young age distribution. The intervention of government in the economy has been used to redistribute income in many countries. The variable GOV attempts to capture this effect. This study also considers the effects of changes over time. The variables $\triangle$AGRI, $\triangle$AGE1, $\triangle$AGE2 and $\triangle$GOV represent absolute changes in these variables over the time periods for which there are data available. (These time periods vary from country to country, depending on the availability of data. See Table 1.)

Since all of these explanatory variables are unitless, it is possible to define them consistently across countries. However, there are some minor differences in definitions. In particular, AGE1 is defined as the fraction of the population under age 15, rather than age 14, in the US and West Germany. Also, GROWTH is defined in terms of Gross Domestic Product for Australia. For West Germany, AGRI is the fraction of employment in agriculture, forestry and fisheries, rather than just agriculture.

2.1 Measures of inequality

For the purposes of any empirical study, it is necessary to summarize the distribution of income into a meaningful index that reflects the level of dispersion of income. There are many different measures of income inequality. One of the most widely used is the Gini concentration ratio, or Gini coefficient. The Gini coefficient can be defined in terms of the average mean difference in the incomes of the recipients:

$$g = \frac{1}{\mu} \int_0^\infty \int_0^\infty |x - y| f(x) f(y) \, dx \, dy, \tag{1}$$

where μ is the mean income of the population, and f( ) represents the density function of income. The Gini coefficient takes on values between 0 (perfect equality) and 1 (perfect inequality).

A graphical device for summarizing the distribution of income is the Lorenz curve. The Lorenz curve compares the cumulative fraction of total income received (on the vertical axis) by the cumulative fraction of recipients (on the horizontal axis), ranked from poorest to richest. For example, a representative point would be that the poorest 10 percent of the population receive 1 per cent of the population's income.

The Lorenz curve suggests another measure of income inequality. In countries with very unequal distributions, the poor will receive a very small portion of the total income. So the fraction of income received by the poorest fifth (or any other quantile) will reflect to some extent the overall distribution of income. Furthermore, this measure emphasizes the status of the relatively poor, a group that deserves special attention. In this paper, I use the fraction of income received by the lowest quintile (Q_1) as another index of income inequality.

Table 2 summarizes all of the data to be used in the analysis. Estimation of the Gini coefficient and the income received by the lowest quintile are discussed below. One interesting aspect of the data is that there has been very little change in income inequality over this period of time. On the other hand, for this sample there has been a substantial growth rate of about 3.5 percent per year. Other variables in the analysis also show moderate variation during the period of the analysis.

2.2 Using grouped data

Almost all data that are readily available on income distributions are reported in some type of grouped format. The most common means of reporting the

Table 2. Summary statistics for data.

Variable	Mean	Standard deviation	Minimum	Maximum
GINI	.3533	.0438	.2783	.4064
$\triangle$GINI	.0003	.0201	−.0356	.0765
Q_1	.0610	.0150	.0406	.0891
$\triangle Q_1$	−.0003	.0055	−.0223	.0076
GROWTH	.0343	.0227	−.0015	.0804
AGRI	.0995	.0985	.0178	.4083
$\triangle$AGRI	−.0289	.0286	−.0944	−.0010
AGE1	.2925	.0544	.2069	.4245
$\triangle$AGE1	−.0120	.0223	−.0513	.0251
AGE2	.0886	.0298	.0280	.1399
$\triangle$AGE2	.0070	.0070	−.0011	.0334
GOV	.2610	.1194	.1322	.6786
$\triangle$GOV	.0150	.0622	−.2000	.1829

Number of observations by country.

Australia	2
Canada	5
Israel	3
Taiwan	4
United Kingdom	3
United States	7
West Germany	3
Total	27

income distribution is to report the actual histogram, e.g. the fraction of recipients whose incomes fall within a specified interval. The sources for the United States, Canada, Israel and Australia take this histogram approach. Another format for reporting income distributions is to report the fraction of total income received by specified quantiles of the population, e.g., the fraction of total income received by each quintile or decile. The sources for Taiwan, United Kingdom and Germany report the income distributions in this format.

Interval data lack information about the distribution of income within the intervals. This poses a problem for calculating measures of income inequality, because there is intra-interval variation in incomes as well as inter-interval variation. Ignoring the intra-group variation will result in an understatement of inequality, so some assumption must be made about the distribution of income within groups. A convenient parametrization of the income distribution has been suggested by Singh and Maddala (1976):

$$f(y) = \frac{a_1 a_2 a_3 y^{a_2-1}}{(1 + a_1 y^{a_2})^{a_3}}, \qquad (2)$$

where y is the level of income and a_1, a_2 and a_3 are parameters. The properties of this density function are further discussed in McDonald and Ransom (1979). McDonald (1984) finds that this density is a better approximation to the size distribution of income than many other 3 and 4 parameters density functions. Once the parameters of the density function are known (or estimated) then the corresponding Gini coefficient can be calculated using the formula reported in McDonald and Ransom (1979):

$$G = 1 - \frac{\Gamma(2a_3 - 1/a_2) \; \Gamma(a_3)}{\Gamma(2a_3) \; \Gamma(a_3 - 1/a_2)}. \qquad (3)$$

The income share of the lowest quintile can be derived from the Lorenz curve. The Lorenz curve corresponding to the Singh-Maddala density function is derived in the Appendix to this paper. Applying those results yields a formula for the income share of the lowest quintile:

$$Q_1 = \frac{a_3 \Gamma(a_3)}{\Gamma(1 + 1/a_2) \; \Gamma(a_3 - 1/a_2)}$$

$$\times \frac{[.8^{-1/a_3} - 1]^{1+1/a_2}}{1 + 1/a_2} \qquad (4)$$

$$\times \; _2F_1[1 + a_3, \frac{a}{a_2} + 1; \frac{a}{a_2} + 2; 1 - .8^{-1/a_3}],$$

where the $_2F_1$ function is the hypergeometric function discussed in McDonald (1984). Note that neither G nor Q_1 depend on the scale parameter a_1.

Estimation of a_2 and a_3 from histogram type data is accomplished by maximizing (numerically) the multinomial log-likelihood function:

$$\ln(L) = \ln(N!) + \sum_{j=1}^{G} [n_j \ln P_j - \ln(n_j!)], \qquad (5)$$

where N is the sample size, n_j is the number of observations in the jth income interval, G is the number of intervals and P_j is the predicted fraction of the population with income in the jth interval. The jth interval is defined as $[y_{j-1}, y_j]$. For the Singh-Maddala density function, the interval fractions

$$P_j = \int_{I_j} f(y) dy$$

have a convenient closed-form representation based on:

$$\int_0^x f(y) dy = 1 - (1 + a_1 x^{a_2})^{-a_3}.$$

See McDonald and Ransom (1979) for a discussion of this estimation method and others applicable to interval data. The estimated values of a_2 and a_3 are used to compute the Gini coefficient and Q_1.

For data in quantile form, a_2 and a_3 are estimated by minimizing the sum-of-squared-errors function

$$SSE = \sum_{j=1}^{5} [F_j - \hat{F}_j]^2, \qquad (6)$$

where F_j is the income share of the jth quintile and $\hat{F}_j$ is the corresponding income share predicted from the Lorenz curve of the Singh-Maddala density function, as derived in the Appendix. The estimated values of a_2 and a_3 are used to calculate the corresponding Gini coefficient. Q_1 is available from the data, of course.

3. Results

The regression analysis of the data is based on the following model:

$$Y_{it} = \beta' X_{it} + \varepsilon_i + \upsilon_{it}, \qquad (7)$$

where i subscripts the country and t the time period. Y is the measure of income inequality, either GINI or Q_1. X_{it} is a vector of explanatory variables for a given country and year. In particular, $GROWTH_{it}$ is defined as the rate of growth in country i during the period between time t and t − 1. β is a vector of parameters to be estimated by the regression. υ_{it} is an error term of the usual type, with mean 0 and constant variance σ^2. ε_i is a country-specific error term which represents differences in the construction of data as well as other country-specific differences in culture, geography, etc., that change very slowly over time, if at all.

A naive way to estimate this model is to ignore ε_i, pooling the data and performing a regression. The results from this analysis are reported in the first two columns of Table 3, under Model Type I. The estimates indicate that higher rates of economic growth are associated with a more equal distribution of income, while increases in the fraction of the population under 14 and the fraction of the population over 65 are associated with a more unequal distribution of income. Higher relative employment in agriculture and higher relative levels of government spending are associated with a less equal distribution of income as measured by the Gini coefficient, but are associated with a relative

Table 3. Pooled cross-section and time series data regression results.*

Model type: Dependent variable:	I GINI	I Q_1	II GINI	II Q_1
Explanatory variable				
Intercept	.0937	.0939	.2918	.0348
	(.1058)	(.0371)	(.0864)	(.0247)
GROWTH	−.4209	.1829	−.1312	.0198
	(.3673)	(.1289)	(.2764)	(.0789)
AGRI	.2031	.0284	.2130	−.0342
	(.1228)	(.0431)	(.1401)	(.0400)
AGE1	.2757	−.0712	.0815	.0227
	(.2227)	(.0781)	(.1659)	(.0474)
AGE2	1.8885	−.3146	.7918	−.0156
	(.4745)	(.1665)	(.5922)	(.1691)
GOV	.0224	.0259	−.0285	.0203
	(.0565)	(.0198)	(.1108)	(.0316)
AUSTRALIA			−.0361	.0136
			(.0170)	(.0049)
CANADA			−.0818	.0266
			(.0136)	(.0039)
ISRAEL			−.0355	.0169
			(.0451)	(.0129)
TAIWAN			−.0999	.0472
			(.0324)	(.0093)
U.K.			−.0190	.0117
			(.0192)	(.0055)
W. GERMANY			−.0275	.0210
			(.0272)	(.0078)
R^2	.6270	.6074	.9102	.9374

* Standard errors of estimates are in parentheses.

improvement of the position of the poorest income groups. None of these effects is measured with much precision, however, and, except for AGE2, none of them would be significantly different from zero using traditional hypothesis testing procedures.

A more reasonable approach to estimating equation (7) is to control for inter-country differences by including a dummy variable for each country, except for the United States. The country variables thus measure differences relative to the United States. The results of the analysis are reported in Table 3, columns 3 and 4, under Model Type II. The most dramatic effect of this specification, relative to Model Type I, is that the coefficients are much smaller, while the standard errors of the estimates are larger. None of the coefficients would be significantly different from zero, with the exception of the country variables.

Another way to eliminate the country-specific effects in equation (7) is to difference the data over time. This leads to a model:

$$Y_{it} - Y_{i,t-1} = \gamma'(X_{it} - X_{i,t-1}) + v_{it} - v_{i,t-1}, \qquad (8)$$

where γ is the parameter vector to be estimated, and the explanatory variables are differenced over time. The results of the analysis are reported in Table 4, as Model Type III. The specification is a bit different here, as GROWTH is not differenced, so it represents the change in the level of economic activity, not the change in the rate of growth. Once again, while the magnitudes are very

Table 4. Time-differenced data regression results.*

Model type: Dependent variable:	III $\triangle$GINI	III $\triangle Q_1$
Explanatory variable		
Intercept	.0106	−.0043
	(.0078)	(.0020)
GROWTH	−.2482	.0822
	(.3032)	(.0790)
$\triangle$AGRI	.0930	−.0172
	(.2210)	(.0576)
$\triangle$AGE1	−.0154	.0209
	(.1943)	(.0506)
$\triangle$AGE2	.1084	.1239
	(.6056)	(.1579)
$\triangle$GOV	−.0009	.0037
	(.0705)	(.0184)
R^2	.1406	.2192

* Standard errors of estimates are in parentheses.

small, and the estimates are measured without much precision, the estimates indicate that economic growth decreases the level of income inequality, and that the poorest groups benefit from economic growth both absolutely and relatively.

One additional fact deserves mention. While the estimated coefficients in these model are small and imprecise, the raw correlation between $\triangle$GINI and GROWTH is quite high, $-.36$. In this analysis I have treated the explanatory variables as exogenous, while they may be determined in part by economic growth. If only growth is the relevant explanatory variable, then the statistical models presented above understate its importance.

4. Conclusion

In this paper I have analyzed the relationship between economic growth and income inequality, using a model that stresses the longitudinal changes within countries, rather than the intercountry differences. The sample is heavily weighted toward highly developed countries, and there has been relatively little change in measures of income inequality over the time period and countries I have examined here. Nevertheless, there is a persistent, though weak, correlation between growth and the measures of income inequality that I have used here. At least in this sample, the poorest income groups benefit most from economic growth, leading to a more equal distribution of income. The effect is small, but is robust with respect to the specification that I have presented in this paper.

5. Acknowledgement

I gratefully acknowledge the helpful research assistance of Edward Jorgensen.

6. Appendix

Derivation of Lorenz curve for the Singh-Maddala density function

It is convenient to express the Lorenz curve corresponding to a particular probability distribution of income as:

$$L(q) = \frac{1}{E(y)} \int_0^q F^{-1}(p) \, dp, \tag{1}$$

where F^{-1} is the inverse of the cumulative distribution function, and $E(y)$ is the expected value of income.

For a technical discussion of the Singh-Maddala function and some of its properties, see McDonald (1984). The cumulative distribution function for the Singh-Maddala distribution can be written as:

$$F(y) = 1 - (1 + a_1 y^{a_2})^{-a_3}.$$

This can be directly inverted:

$$F^{-1}(p) = \left[\left(\frac{1}{a_1}\right)((1-p)^{-1/a_3} - 1)\right]^{1/a_2} \tag{2}$$

Also note that

$$E(y) = \frac{\Gamma(1 + 1/a_2)\,\Gamma(a_3 - 1/a_2)}{a_1^{1/a_2}\,\Gamma(a_3)}. \tag{3}$$

The first step in deriving the Lorenz curve is to evaluate

$$\int_0^q F^{-1}(p)\,dp = \int_0^q \left[\left(\frac{1}{a_1}\right)((1-p)^{-1/a_3} - 1)\right]^{1/a_2} dp. \tag{4}$$

By substituting $s = [(1-p)^{(-1/a_3)} - 1]$ expression (4) can be written as

$$\int_0^{(1-q)^{-1/a_3}-1} (1/a_1)^{1/a_2}\, s^{1/a_2}\, [a_3(s+1)^{-a_3-1}]\ ds.$$

From Gradshteyn and Ryzhik (1965), page 284, equation 3.194, this can be evaluated as

$$a_3\left(\frac{1}{a_1}\right)^{1/a_2}\left[\frac{[(1-q)^{-1/a_3} - 1]^{1+1/a_2}}{1 + 1/a_2}\right]$$

$$\times\ {}_2F_1[1 + a_3, 1 + 1/a_2; 2 + 1/a_2; 1 - (1-q)^{-1/a_3}],$$

where the ${}_2F_1$ function is the hypergeometric function. See McDonald (1984) for a discussion of this function. Dividing by equation (3), $E(y)$, yields the desired result:

$$L(q) = \frac{a_3\Gamma(a_3)}{\Gamma(1 + 1/a_2)\,\Gamma(a_3 - 1/a_2)}\,\frac{[(1-q)^{-1/a_3} - 1]^{1+1/a_2}}{1 + 1/a_2}$$

$$\times\ {}_2F_1[1 + a_3, 1 + 1/a_2; 2 + 1/a_2; 1 - (1-q)^{-1/a_3}].$$

Bibliography

Ahluwalia MS. 1974. Income inequality: some dimensions of the problem, chapter 1 in Chenery: *Redistribution with growth*. London: Oxford University Press.

Berry A. 1974. Changing income distribution under development: Columbia. *Review of Income and Wealth*, 20: 289–316.

Gradshteyn IS and Ryzhik IM. 1965. *Tables of integrals, series and products*. New York: Academic Press.

Kuznets S. 1963. Quantitative aspects of the economic growth of nations: VIII. Distribution of Income by Size, *Economic Development and Cultural Change*, 11: 1–80.

McDonald JB and Ransom MR. 1979. Functional forms, estimation techniques and the distribution of income. *Econometrica*, 47: 1513–1526.

McDonald JB. 1984. Some generalized functions for the size distribution of income. *Econometrica*, 52: 647–663.

Oshima HT. 1962. The international comparison of size distribution of family incomes with special reference to Asia. *The Review of Economics and Statistics*, 44: 439–445.

Singh SK and Maddala GS. 1976. A function for the size distribution of incomes. *Econometrica*, 44: 963–970.

Soltow L. 1968. Long run changes in British income inequality. *Economic History Review*, 21: 17–29.

Weiskoff R. 1970. Income distribution and economic growth in Puerto Rico, Argentina and Mexico. *Review of Income and Wealth*, 16: 303–332.

D. Methodology

The coefficient of determination for regression without a constant term

ANTON P. BARTEN

1. Introduction

R^2, the coefficient of determination or the squared correlation coefficient, is a recurrent theme in statistics. Kendall (1960) calls it an evergreen. Sooner or later any empirical analyst has to deal with some aspect of R^2 which appears not to have been treated satisfactorily in the known literature. Some like Hotelling (cf. Kendall (1960)) or Cramer (1964, 1984) even do it sooner *and* later. Also the present contribution is prompted by the problem that the usual expressions for R^2 may yield unacceptable values in regressions without a constant term. The purpose is to obtain an expression which avoids this problem and at the same time is as analogous as possible for regressions with and without intercept.

Regressions without an intercept are no rarity in applied econometrics. As explained e.g. in Cramer (1969) one of the set of hypotheses which constitute the theory of the consumption function of Friedman (1957) is the proportionality of consumption (C) and permanent income (Y_p):

$$C = kY_p$$

a relationship obviously without intercept. Another example is the case of weighted regression. Transforming the variables to obtain homoskedastic disturbances implies changing the dummy constant into a true variable. The resulting equation has then no constant term. In regressions with first differences of time series a constant term represents a trend in the level values of the dependent variables. Such a trend may be a nuisance, economically meaningless and posing problems for extrapolations. One then rightly omits the constant term from the original regression.

The commonly used expressions for R^2 are somewhat awkward in regressions without an intercept. They can give negative values for R^2 or values larger than unity. One approach, taken by the TSP regression package, for example, is not to report any value at all. Applied workers quickly find out that one cannot get rid of the problem in this way. The people who commission the project or the referees of the journal to which the article is submitted know usually enough of the field to demand an R^2 and not enough to know that the coefficient of determination has many defects as a measure of goodness of fit and that other measures are perhaps more suitable. The econometric literature, as far as consulted by the author, is silent on this score, except for one source: Theil (1971) – of course, one is tempted to say – where in a footnote two suggestions for an R^2 in the case of regressions without constant term are made. These will be reviewed at a later stage. They differ from the expression proposed here which is basically identical for regressions with and without constant term and which satisfies a set of conventional properties.

The desired analogy with R^2 in case of regressions with constant term makes it natural to start off in the next section with the usual theory about this concept. One problem with a discussion of R^2 is that in the general linear model its counterpart in the population is not specified.

Section 3 defines and discusses a parent concept for R^2 and summarizes the various properties of R^2 (in regression with a constant term) which one would like to see satisfied also by R^2 in regressions without intercept. Section 4 deals with such regressions, while Section 5 discusses the merits of various measures of determination or of goodness of fit in that case. Conclusions form the ending Section.

2. Some theory about regressions without constant term

The regression model considered here is written as

$$y = X\gamma + u \tag{2.1}$$

where y is a T-vector of (potential) observations on a dependent variable, X is a T×k matrix of observations on *nonrandom* explanatory variables, γ is a k-vector of unknown coefficients. The T-vector u is unobserved. It is assumed to be distributed with $E(u) = 0$ and $E(uu') = \sigma^2 I$ (σ^2 a positive finite scalar). The assumption of such a scalar covariance matrix is somewhat more specific than is actually needed. A generalization, however, is rather obvious.

To make (2.1) relevant for the discussion of the present section it is specified that the first column of the matrix X consists of ones:

$$X = (\iota, Z) \tag{2.2}$$

where ι is a T-vector with all elements equal to one, while Z is the T × (k − 1) matrix of all other columns of X. Correspondingly, γ will be decomposed as:

$$\gamma = \begin{pmatrix} \alpha \\ \beta \end{pmatrix} \tag{2.3}$$

where α denotes the intercept and β the other $k-1$ regression coefficients. Thus (2.1) can be rewritten as:

$$y = (\iota, Z) \begin{pmatrix} \alpha \\ \beta \end{pmatrix} + u = \alpha\iota + Z\beta + u. \tag{2.4}$$

It will be assumed throughout that the matrix X has full column rank, that for $T \to \infty$ lim $(1/T)$ X'X exists and is nonsingular.

Under the assumptions made the best linear unbiased and consistent estimator of γ in (2.1) is

$$c = (X'X)^{-1} X'y \tag{2.5}$$

which, as can be readily verified, is decomposable as:

$$c = \begin{pmatrix} a \\ b \end{pmatrix} = \begin{pmatrix} \bar{y} - \bar{z}'(Z'NZ)^{-1} Z'Ny \\ (Z'NZ)^{-1} Z'Ny \end{pmatrix} \tag{2.6}$$

with $\bar{y} = (1/T)\iota'y$, $\bar{z}' = (1/T)\iota'Z$ and

$$N = I - (1/t)\iota\iota' = I - \iota(\iota'\iota)^{-1}\iota', \tag{2.7}$$

a notation borrowed from Kloek (1961). The matrix N is a symmetric idempotent matrix which represents the operation of taking deviations from the column (or row) mean like:

$$NZ = Z - \iota\bar{z}', \quad Ny = y - \iota\bar{y}.$$

The matrix Z'NZ plays an important rôle in what follows. It is then useful to note that it can be obtained from X'X by premultiplying that matrix by the matrix $(-\bar{z}, I)$ and postmultiplying it by the transpose of the latter matrix. Since $(-\bar{z}, I)$ has full row rank and X'X has full rank Z'NZ has full rank too. This also holds for $M = (1/T)$ Z'NZ. It follows from the limit properties of $(1/T)$X'X that $M_L = \lim M$ for $T \to \infty$ exists and is nonsingular.

There is usually no reason to treat the intercept differently from the other regression coefficients, except precisely in the case of the coefficient of determination. This coefficient is defined as:

$$R^2 = \frac{b'Z'NZb}{y'Ny} \tag{2.8}$$

or, with M as defined above and $m = (1/T)$ y'Ny, as

$$R^2 = b'Mb/m. \tag{2.9}$$

Another expression for R^2 involves the T-vector of residuals. This is defined as:

$$\hat{u} = y - Xc. \tag{2.10}$$

It follows from (2.5) that $X'\hat{u} = 0$. Consequently, $\iota'\hat{u} = 0$ and

$$N\hat{u} = \hat{u}. \tag{2.11}$$

Noting that $NXc = NZb$ one can write

$$\begin{aligned} y'Ny &= b'Z'NZb + 2\hat{u}'NZb + \hat{u}'N\hat{u} \\ &= b'Z'NZb + \hat{u}'\hat{u} \end{aligned} \tag{2.12}$$

since $X'\hat{u} = 0$ also implies $Z'\hat{u} = Z'N\hat{u} = 0$. This result is used to rewrite (2.8) as

$$R^2 = 1 - \frac{\hat{u}'\hat{u}}{y'Ny} \tag{2.13}$$

or with $h = (1/T)\hat{u}'\hat{u}$ as

$$R^2 = 1 - h/m. \tag{2.14}$$

One may consider $b'ZNZb$ in decomposition (2.12) the 'explained' part of the sample variation in the dependent variable. The coefficient of determination R^2 is then interpreted as the fraction of the sample variation in the dependent variable explained by the regressors. As is clear from (2.8) R^2 is positive while (2.13) shows that it is not larger than unity.

3. Parent concept

The coefficient R^2 is a measure of goodness of fit, but what does it actually measure? Does it capture something more than a purely incidental co-relation, unique for a particular sample? In bivariate analysis the sample correlation coefficient is an estimator of:

$$\varrho = \frac{\sigma_{12}}{\sqrt{\sigma_{11} \sigma_{22}}} \tag{3.1}$$

where σ_{11} and σ_{22} are the population variances of variables 1 and 2, respectively, and σ_{12} is the population covariance between those two variables. Its square is given by

$$\varrho^2 = \frac{\sigma_{12}^2}{\sigma_{11} \sigma_{22}} = \frac{\zeta \sigma_{22} \zeta}{\sigma_{11}}$$

with $\zeta = \sigma_{12}/\sigma_{22}$. This concept can be readily extended to the multivariate case:

$$\zeta' \Sigma_{22} \zeta / \sigma_{11} \tag{3.2}$$

with $\zeta = \Sigma_{22}^{-1} \Sigma_{21}$, Σ_{22} being the covariance matrix of the random regressors and Σ_{21} the covariance matrix of the regressand with the regressors.

In regression model (2.1) the regressors are nonrandom and their covariance with the (random) regressand is zero. However, an expression like (2.9) is zero only with measure zero and it is not acceptable to have then zero as its counterpart in the population. An alternative is to follow Cramer (1984) and use the value to which the random variable R^2 converges in probability for $T \to \infty$ as the parent concept. Under the assumptions made

$$\text{plim } b'Mb = \beta'M_L\beta \tag{3.3}$$

$$\begin{aligned}\text{plim } m &= \text{plim } b'Mb + \text{plim } h \\ &= \beta'M_L\beta + \sigma^2.\end{aligned} \tag{3.4}$$

Thus

$$\text{plim } R^2 = \beta'M_L\beta/(\beta'M_L\beta + \sigma^2) = P^2 \tag{3.5}$$

is the parent counterpart of R^2. Given that the regressors are nonrandom and in principle under control of the researcher one may view M_L as a population characteristic – see also Koerts and Abrahamse (1970) on this issue. Then P^2 may be considered as the fraction of the population variance of the dependent variable explained by the variation of the regressors.

Equation (3.5) suggests another expression for R^2, namely

$$R^2 = \frac{b'Z'NZb}{b'Z'NZb + \hat{u}'\hat{u}} = \frac{b'Mb}{b'Mb + h}, \tag{3.6}$$

from which it is clear that $R^2 = 0$ iff $b = 0$ and $R^2 = 1$ iff $\hat{u} = 0$.

The discussion until this far has been in terms of a regression with an intercept. The alternative expressions for R^2 all have the properties
(i) plim $R^2 = P^2$ for $T \to \infty$
(ii) $0 \le R^2 \le 1$
(iii) $b = 0 \leftrightarrow R^2 = 0$
(iv) $\hat{u} = 0 \leftrightarrow R^2 = 1$.

The last two represent very much the ideal of R^2 as a measure of the goodness of fit. If the Z variables together have no explanatory power ($b = 0$) then $R^2 = 0$, if they describe the dependent variable completely ($\hat{u} = 0$) then $R^2 = 1$. Properties (i) through (iv) are the ones which one would like to see satisfied also by the coefficient of determination for regressions without intercept. To these regressions we turn in the next section.

4. Regressions without constant term

Formally (2.1) and (2.4) still hold, except that α is set equal to zero:

$$y = Z\beta + u \tag{4.1}$$

where u has the properties given in Section 2. The least-squares estimator

$$b_c = (Z'Z)^{-1} Z'y \tag{4.2}$$

is the best linear unbiased and consistent estimator of β in (4.1). It can be directly compared with its counterpart (2.6). Define

$$\hat{u}_c = y - Zb_c \tag{4.3}$$

as the residual vector corresponding to (4.1) and (4.2). It is simple to derive that $Z'\hat{u}_c = 0$. In this case, however, the residuals do not identically add up to zero. Let $\bar{y}$ and $\bar{z}'$ be the sample averages of y and the columns of Z, respectively. Then

$$\bar{u}_c = (1/T)\iota'\hat{u}_c = \bar{y} - \bar{z}'b_c \tag{4.4}$$

is the average residual.

Note that under the assumptions made $E(\bar{u}_c) = 0$ and

$$E(\bar{u}_c^2) = \sigma^2 \{1 - z'[(1/T)Z'Z]^{-1} z/T$$

which for $T \to \infty$ converges to zero. This means that $\bar{u}_c$ converges in the quadratic mean to zero, so that for $T \to \infty$ also

$$\text{plim } \bar{u}_c = 0. \tag{4.5}$$

For small samples, however, $\bar{u}_c$ is usually not equal to zero.

In regressions with constant term the equivalence of R^2 expressions (2.8), (2.9), (2.13), (2.14) and (3.6) followed from (2.12). The counterpart of the relation in the present case is:

$$y'Ny = b_c' Z'NZb_c + 2\hat{u}_c'NZb_c + \hat{u}_c'N\hat{u}_c$$
$$= b_c' Z'NZb_c + 2\hat{u}_c' Zb_c - 2\bar{u}_c\iota'Zb_c + \hat{u}_c'\hat{u}_c - T\bar{u}_c^2 \tag{4.6}$$
$$= b_c' Z'NZb_c + \hat{u}_c'\hat{u}_c - T(2\bar{u}_c \bar{z}'b_c + \bar{u}_c^2)$$

from which it follows that

$$m = b_c' Mb_c + (1/T)\hat{u}_c'\hat{u}_c - (2\bar{u}_c\bar{z}' b_c + \bar{u}_c^2). \tag{4.7}$$

If is not difficult to verify that

$$\text{plim } m = \beta'M_L\beta + \sigma^2, \tag{4.8}$$

which corresponds to (3.4) for regressions with an intercept. For small samples the last term in (4.6) or in (4.7) causes problems when using conventional R^2 expressions.

Consider, for example, (2.8) with b replaced by b_c:

$$R_1^2 = b_c' Z'NZb_c/(y'Ny). \tag{4.9}$$

Although plim $R_1^2 = P^2$, $R_1^2 = 0$ if $b_c = c\,0$, $R_1^2 = 1$ if $\hat{u}_c = 0$, there is no fixed upper bound to R^2 because the presence of the last term in (4.6) may cause $y'Ny$ to be smaller than $b_c'Z'NZb_c$. Clearly, values of $R^2>1$ will raise eyebrows with some.

Turn next to (2.13), with $\hat{u}$ replaced by $\hat{u}_c$:

$$R_2^2 = 1 - \hat{u}_c'\hat{u}_c/y'Ny \tag{4.10}$$

which is one of the two measures proposed by Theil (1971, p. 178). Also plim $R_2^2 = P^2$. Clearly, R_2^2 is not larger than one, and $R^2 = 1$ if $\hat{u}_c = 0$. However, $b_c = 0$ does not mean $R_2^2 = 0$. In fact R_2^2 can take on negative values because there is no constraint on the sign of

$$y'Ny - \hat{u}_c'\hat{u}_c = b_c'\,Z'NZb_c - T(2\bar{u}_c\bar{z}'b_c + \bar{u}_c^2).$$

One can, of course, amend (4.10) to read

$$R_3^2 = 1 - \hat{u}_c'N\hat{u}_c/y'Ny \tag{4.11}$$

which would still be analogous to (2.13) because of (2.11). Also this expression converges in probability to P^2 and is not larger than one. Its upper bound is reached if $N\hat{u}_c = 0$, which is a weaker condition than $\hat{u}_c = 0$. If $b_c = 0$, $\hat{u}_c = y$ and thus $R^2 = 0$. Still R^2 can be negative since the sign of

$$y'Ny - \hat{u}_c'N\hat{u}_c = b_c'Z'NZb_c - 2T\bar{u}_c\,\bar{z}'b_c$$

is not determined. Note that

$$R_3^2 - R_2^2 = \bar{u}_c^2/m > 0,$$

i.e. R_3^2 will usually be larger than R_2^2. As Theil points out this pleads against the use of R_3^2 since a nonzero average residual detracts from the goodness of fit and should thus be taken into account.

Next to (4.10) Theil suggests another alternative:

$$R_4^2 = 1 - \hat{u}_c'\hat{u}_c/y'y. \tag{4.12}$$

This expression meets properties (ii) through (iv) but not (i). In fact, for $T \to \infty$ and $\bar{z}_L = \lim \bar{z}$

$$\text{plim } R_4^2 = \frac{\beta'M_L\beta + (\beta'\bar{z}_L)^2}{\beta'M_L\beta + (\beta'\bar{z}_L)^2 + \sigma^2} \geq P^2.$$

Moreover, the use of $y'y$ rather than $y'Ny$ departs from the analogy with the case of regressions with constant term.

As a fifth alternative consider expression (3.6) with b replaced by b_c and $\hat{u}$ by $\hat{u}_c$:

$$R_5^2 = \frac{b_c'\,Z'NZb_c}{b_c'\,Z'NZB_c + \hat{u}_c'\hat{u}_c} \tag{4.13}$$

It meets all properties of Section 3. Nonzero average residuals increase the denominator and in this way reduce the goodness-of-fit measure. An alternative of R_5^2 with $\hat{u}_c'\hat{u}_c$ replaced by $\hat{u}_c'N\hat{u}_c$ would also meet the properties of Section 3 but would not reflect the nonzero average residual effect.

It appears that R^2 is superior to all the other expressions considered in this section. There is one minor blemish, however. Its denominator will not stay the same if the right-hand side of (4.1) is changed, for example, because of imposing restrictions on the β vector. Consequently, one should be careful in using R^2 to compare the empirical performance of different regressions with the same dependent variable. For the same reason it cannot be used in the usual F-statistic for testing coefficient constraints in the same way as R^2 (in case of regressions with a constant term) can be employed.

5. Conclusions

It is proposed to use a general expression for the determination coefficient

$$R^2 = \frac{b_*'Z'NZb_*}{b_*'Z'NZb_* + \hat{u}_*'\hat{u}_*} \tag{5.1}$$

where b_* is the vector of estimated regression coefficients of the explanatory variables (excluding the dummy constant, if that is specified) and u_* is the vector of residuals defined as

$$\hat{u}_* = N(y - Zb_*)$$

in the case of regressions with constant term and as

$$\hat{u}_* = y - Zb_*$$

in the case of regression without constant term. An alternative and sometimes more convenient expression is to write

$$R^2 = \frac{\hat{y}'N\hat{y}}{\hat{y}'N\hat{y} + \hat{u}_*'\hat{u}_*} \tag{5.2}$$

where

$$\hat{y} = Zb_*.$$

Expressions (5.1) and (5.2) have the same properties for regressions with and without constant term. In the former case they are equivalent to the traditional definitions (2.8) or (2.13).

Finally, it is pointed out that the properties of R^2 hold for b_* being any consistent estimator of β. Expressions (5.1) or (5.2) can also be applied in multivariate problems where a set of regression equations is estimated simultaneously and one would like to know the goodness of fit for each equation separately.

6. Acknowledgement

The author thanks Mr. Denis de Crombrugghe for his remarks on an earlier version of this paper. He stays responsible for any remaining errors.

Bibliography

Cramer JS. 1964. Efficient grouping, regression and correlation in Engel curve analysis. *Journal of the American Statistical Association*, 59: 233–250.

Cramer JS. 1969. *Empirical Econometrics*. Amsterdam: North-Holland Publishing Co.

Cramer JS. 1984. *Sample size and R^2*. Discussion paper AE N2/84 of Faculty of Actuarial Science and Econometrics of the University of Amsterdam.

Friedman M. 1957. *A theory of the consumption function*. Princeton: Princeton University Press.

Kendall MG. 1960. The evergreen correlation coefficient, in: *Contributions to probability and statistics. Essays in Honor of Harold Hotelling*, edited by I. Olkin a.o. Stanford: Stanford University Press, 274–277.

Kloek T. 1961. Convenient matrix notations in statistics and aggregation theory. *International Economic Review*, 2: 351–360.

Koerts J and Abrahamse APJ. 1970. The correlation coefficient in the general linear model. *European Economic Review*, 1: 401–427.

Theil H. 1971. *Principles of Econometrics*. Amsterdam: North-Holland Publishing Co.

The coefficient of determination revisited

RISTO D.H. HEIJMANS and HEINZ NEUDECKER

1.

The coefficient of determination (R^2) has received much attention as a goodness-of-fit measure for regressions. Many researchers use R^2 as a sort of magical pointer, they will favour the model which engenders the higher coefficient of determination. Other investigators ridicule this approach and raise the objection that R^2 is a random variable and as such can be near unity while the model under consideration is worthless; the reverse obviously can also be true.

So the question arises about the properties of R^2 (if any), and this the authors try to answer. We will show that R^2 can be considered a consistent estimator of some rather complicated function of the parameters of the model under examination, when certain assumptions relating to the matrix X hold. This is also the approach of Koerts and Abrahamse (1970), Barten (1987) and Cramer (1987) and we will pursue this particular avenue somewhat further.

2.

It seems to be common practice to define R^2 as the ratio of explained and total variation. This definition, however, seems to break down in the case of a regression without an intercept. (See Barten (1987) for more details.)

The present authors will suggest another procedure that can handle all cases, viz. regression with or without an intercept, non-scalar dispersion matrices etc. In the case of an intercept and a scalar dispersion matrix our definition will lead to the usual R^2.

We will start with the following well-known model

$$y = X\beta + \varepsilon, \tag{2.1}$$

where y and ε are (n × 1) vectors, the components of ε viz. $\varepsilon_1, \ldots, \varepsilon_n$ are i.i.d.

R.D.H. Heijmans and H. Neudecker (eds.), The Practice of Econometrics. ISBN 90-247-3502-5.
© 1987, Martinus Nijhoff Publishers, Dordrecht. Printed in the Netherlands.

with $E\varepsilon_i = 0$ and $E\varepsilon\varepsilon' = \sigma^2 I$, X is a nonstochastic (n × k) matrix of rank k<n (k fixed) and β is a (k × 1) vector of unknown constants.
Further

$$s := (1, \ldots, 1)'; \ N_s := I - (s's)^{-1}ss'; \ N_x := I - X(X'X)^{-1}X', \qquad (2.2)$$

where := is used to define the left-hand side in terms of the right-hand side. The matrices N_s and N_x are symmetric idempotent and in the case that one column of X equals s then $N_s N_x = N_x N_s = N_x$. The following definitions and properties obtain:

$$b := (X'X)^{-1} X'y = \beta + (X'X)^{-1} X'\varepsilon \qquad (2.3)$$

$$\hat{y} = Xb, \ e := y - \hat{y} = N_x \varepsilon. \qquad (2.4)$$

Clearly

$$X'e = 0 \text{ and } \hat{y}'e = 0. \qquad (2.5)$$

We now define R^2 as follows:

$$R^2 := \frac{(\hat{y}'N_s y)^2}{\hat{y}'N_s \hat{y} \cdot y'N_s y}. \qquad (2.6)$$

This is in fact the squared sample correlation coefficient between y and ŷ. In the case that X has a column s, the R^2 of (2.6) will coincide with the usual R^2. Let us investigate the asymptotic properties of the statistic R^2 of (2.6) which can be considered to be the estimator of a sort of population correlation coefficient. As most of our proofs rely heavily on certain properties of eigenvalues, we will first prove a useful lemma about eigenvalues.

Lemma 1

If there exist m, M with $0 < m < M < \infty$ such that

$$m \leq \lambda_{1n} \leq \ldots \leq \lambda_{kn} \leq M$$

then

$$0 < \mu_{2n} \leq \ldots \leq \mu_{kn} < \infty, \text{ where } \lambda_{1n} \leq \ldots \leq \lambda_{kn} \text{ are}$$

the eigenvalues of $\frac{1}{n} X'X$ and $\mu_{1n} \leq \ldots \leq \mu_{kn}$ are

the eigenvalues of $\frac{1}{n} X'N_s X$.

Proof

We will use Poincaré's Separation Theorem: If $A = A'$ ($n \times n$) with eigenvalues $\lambda_1 \leq \ldots \leq \lambda_n$ and G ($n \times k$) is such that $G'G = I_k$, then the eigenvalues of $G'AG$

$$\mu_1 \leq \ldots \leq \mu_k \text{ obey}$$

$$\lambda_i \leq \mu_i \leq \ldots \leq \lambda_{n-k+i}. \ (i = 1, \ldots, k)$$

In our case

$$\frac{1}{n} X'N_s X = \frac{1}{n} X'TT'X \text{ where } N_s = TT' \text{ and }$$

$$T'T = I_{n-1}.$$

For the eigenvalues of $\frac{1}{n} T'XX'T$, $\alpha_{1n} \leq \ldots \leq \alpha_{n-1,n}$ we clearly have that $\alpha_{1n} = \ldots = \alpha_{n-k-1,n} = 0$, and of course $\alpha_{n-k,n} = \mu_{1n}, \ldots, \alpha_{n-1,n} = \mu_{kn}$. Further $\frac{1}{n} XX'$ has eigenvalues $\xi_{1n} \leq \ldots \leq \xi_{nn}$, with $\xi_{1n} = \ldots = \xi_{n-k,n} = 0$, $\xi_{n-k+1,n} > 0$ and $\xi_{n-k+1,n} = \lambda_{1n}, \ldots, \xi_{nn} = \lambda_{kn}$. Now according to Poincaré's Theorem we have

$$\xi_{in} \leq \alpha_{in} \leq \xi_{i+1,n} \ (i = 1 \ldots n-1).$$

So obviously

$$0 < \xi_{n-k+1,n} = \lambda_{1n} \leq \alpha_{n-k+1,n} = \mu_{2n} \leq \xi_{n-k+2,n} = \lambda_{2n}.$$

Given the conditions of the lemma it is clear that

$$0 < \mu_{2n} < \infty.$$

By the same reasoning we get

$$0 < \xi_{n-1,n} = \lambda_{k-1,n} \leq \alpha_{n-1,n} = \mu_{kn} \leq \xi_{nn} = \lambda_{kn}, \text{ hence}$$

$$0 < \mu_{kn} < \infty.$$

What we actually have shown is that if $\frac{1}{n} X'X$ remains a finite matrix of rank k, then $\frac{1}{n} X'N_s X$ remains a finite matrix of rank $\geq k-1$ (k-1 if $\mu_{1n} = 0$ and k if $\mu_{1n} > 0$). Before we investigate the asymptotic properties of R^2 we will state a lemma dealing with asymptotic properties of uniformly bounded random variables having a probability limit.

Lemma 2

If the sequence (z_n) of random variables is uniformly bounded then plim $z_n = z$ implies

$$E|z_n - z|^r \to 0 \text{ as } n \to \infty \text{ for all } r > 0.$$

Proof: see Lukacs (1975, p. 38).
Now we are able to prove our first result.

Lemma 3

If the conditions under (2.1) obtain and if in addition the assumption of Lemma 1 holds, we have

(i) $\text{Plim } \dfrac{1}{n} (\hat{y}'N_s y - \beta'X'N_s X\beta) = 0$

(ii) $\text{Plim } \dfrac{1}{n} (\hat{y}'N_s \hat{y} - \beta'X'N_s X\beta) = 0$

(iii) $\text{Plim } \dfrac{1}{n} (y'N_s y - \beta'X'N_s X\beta - n\sigma^2) = 0.$

If we also assume that $\lim\limits_{n\to\infty} \dfrac{1}{n} X'N_s X = H$, then we have

$$\text{Plim } R^2 = \text{Plim } (\hat{y}'N_s y)^2 / (\hat{y}'N_s \hat{y} \cdot y'N_s y) = (\beta'H\beta)/(\beta'H\beta + \sigma^2).$$

Further $\lim\limits_{n\to\infty} ER^2 = \beta'H\beta/\beta'H\beta+\sigma^2)$ and $\lim\limits_{n\to\infty} D(R^2) = 0$. Hence Plim $R^2 = 1$ iff $\sigma^2 = 0$

Proof

$$\hat{y} = X\beta + (X'X)^{-1}X'\varepsilon, \text{ hence}$$

$$\dfrac{1}{n} \hat{y}'N_s y = \dfrac{1}{n} \{\beta'X' + \varepsilon'X(X'X)^{-1}X'\} N_s (X\beta + \varepsilon)$$

$$\to \dfrac{1}{n} (\hat{y}'N_x y - \beta'X'N_s X\beta) = \dfrac{1}{n} \beta'X'N_s\varepsilon + \dfrac{1}{n} \beta'X'N_s X(X'X)^{-1}X'\varepsilon$$

$$+ \dfrac{1}{n} \varepsilon'X(X'X)^{-1}X'N_s\varepsilon.$$

We have $D(\dfrac{1}{n}\beta'X'N_s\varepsilon) = \dfrac{1}{n^2}\sigma^2\beta'X'N_s X\beta \leq \dfrac{1}{n}\sigma^2\mu_{kn}\beta'\beta$. This last expression tends to zero as $n \to \infty$. Let us now turn to the second term of the right-hand side. We get

$$D\{\frac{1}{n}\beta'X'N_sX\,(X'X)^{-1}\,X'\varepsilon\} = \frac{1}{n^2}\sigma^2\beta'X'N_sX(X'X)^{-1}\,X'N_sX\beta$$

$$\leq \frac{1}{n^2}\sigma^2\,\beta'X'N_sX\beta,$$

for the greatest eigenvalue of $X(X'X)^{-1}X'$ is 1. So this expression will also tend to zero if $n \to \infty$. Finally $\frac{1}{n}\varepsilon'X(X'X)^{-1}X'N_s\varepsilon = (\frac{1}{n}\varepsilon'X)(\frac{1}{n}X'X)^{-1}\cdot(\frac{1}{n}X'N_s\varepsilon)$, where

$$D(\frac{1}{n}X'\varepsilon) = \frac{1}{n^2}\sigma^2\,X'X \text{ and } D(\frac{1}{n}X'N_s\varepsilon) = \frac{1}{n^2}\sigma^2 X'N_sX.$$

Both variance matrices tend to zero if $n \to \infty$. This concludes the proof of assertion (i). Proceeding in the same manner one can write

$$\frac{1}{n}\hat{y}'N_s\hat{y} = \frac{1}{n}\{\beta'X' + \varepsilon'X\,(X'X)^{-1}X'\}\,N_s\,\{X\beta + X(X'X)^{-1}X'\varepsilon\},$$

which yields

$$\frac{1}{n}(\hat{y}'N_s\hat{y} - \beta'X'N_sX\beta) =$$

$$\frac{2}{n}\beta'X'N_sX\,(X'X)^{-1}X'\varepsilon + \frac{1}{n}\varepsilon'X(X'X)^{-1}\,X'N_sX\,(X'X)^{-1}X'\varepsilon.$$

The variance of the first right-hand side term goes to zero as $n \to \infty$ as already shown in the first part of the proof. We now turn to the second term:

$$\frac{1}{n}\varepsilon'X\,(X'X)^{-1}\,X'N_sX\,(X'X)^{-1}X'\varepsilon =$$

$$(\frac{1}{n}\varepsilon'X)(\frac{1}{n}X'X)^{-1}(\frac{1}{n}X'N_sX)(\frac{1}{n}X'X)^{-1}(\frac{1}{n}X'\varepsilon)$$

and because $D(\frac{1}{n}X'\varepsilon)$ tends to zero if $n \to \infty$ we have proven assertion (ii). Ultimately

$$\frac{1}{n}y'N_sy = \frac{1}{n}(\beta'X' + \varepsilon')N_s(X\beta + \varepsilon) =$$

$$\frac{1}{n}\beta'X'N_sX\beta + \frac{2}{n}\beta'X'N_s\varepsilon + \frac{1}{n}\varepsilon'N_s\varepsilon.$$

This yields

$$\frac{1}{n}(y'N_sy - \beta'X'NX\beta - n\sigma^2) = \frac{2}{n}\beta'X'N_s\varepsilon + \frac{1}{n}\varepsilon'N_s\varepsilon - \sigma^2.$$

Now $D(\frac{1}{n} \beta'X'N_s\varepsilon) = \frac{\sigma^2}{n^2} \beta'X'N_sX\beta$ which tends to zero as $n \to \infty$. Further $\frac{1}{n} \varepsilon'N_s\varepsilon = \frac{1}{n} \varepsilon'\varepsilon - (\frac{1}{n} \varepsilon's)^2$. Clearly $\text{Plim}(\frac{1}{n} \varepsilon's) = 0$, as $D(\frac{1}{n} \varepsilon's)$ tends to zero if $n \to \infty$. By Khintchine's law $\text{Plim} \frac{1}{n} \varepsilon'\varepsilon = \sigma^2$. This concludes the proof of assertion (iii) and the first part of Lemma 3. The second part follows immediately. The above lemma shows that R^2 is a consistent estimator of the reciprocal of $1 + \sigma^2/\beta'H\beta$, when the assumption pertaining to $\lim_{n \to \infty} \frac{1}{n}$ X'N$_s$X holds. We see that the ratio $\sigma^2/\beta'H\beta$ is then of vital importance. To clarify the mystery of the intercept we will rewrite (2.6) as

$$R^2 = (\hat{y}'N_s\hat{y} - n\hat{\bar{y}}\bar{e})^2 / \{\hat{y}'N_s\hat{y}(\hat{y}'N_s\hat{y} + e'N_se - 2n\hat{\bar{y}}\bar{e})\} \tag{2.7}$$

where $\hat{\bar{y}} := \frac{1}{n} s'\hat{y}$ and $\bar{e} := \frac{1}{n} s'e$. In the derivation of (2.7) we have substituted $\hat{y} + e$ for y in formula (2.6). In the case of a constant term, where $\bar{e} = 0$, we arrive at

$$R^2 = \hat{y}'N_s\hat{y} / (\hat{y}'N_s\hat{y} + e'e). \tag{2.8}$$

We now turn to the model where the elements of ε are no longer i.i.d., but obey $E\varepsilon = 0$ and $E\varepsilon\varepsilon' = \sigma^2 V$ (V a p.d. matrix). For this model we make the following assumptions

$$0 < m \leq \lambda_{1n} \leq \ldots \leq \lambda_{kn} \leq M < \infty \text{ and} \tag{2.9}$$

$$0 < m \leq \mu_{1n} \leq \ldots \leq \mu_{nn} \leq M < \infty, \tag{2.10}$$

where $\{\lambda_{in}\}_{i=1}^{k}$ and $\{\mu_{in}\}_{i=1}^{n}$ are the eigenvalues of $\frac{1}{n}$ X'X and V^{-1} respectively.

$$\text{Plim}(\frac{1}{n} \varepsilon'\varepsilon - \frac{\sigma^2}{n} \text{tr}V) = 0. \tag{2.11}$$

(Note that if ε is normally distributed (2.10) implies (2.11).) The model produces the well-known GLS estimator

$$b = (X'V^{-1}X)^{-1} X'V^{-1}y. \tag{2.12}$$

Of course one can rewrite the model as

$$z = W\beta + u, \tag{2.13}$$

where $z := V^{-1/2}y$; $W := V^{-1/2}X$ and $u := V^{-1/2}\varepsilon$. Obviously if the original model has an intercept, the derived model in general will be devoid of a constant term. The problem now arises of choosing R^2, for we can either base it on the original model or on (2.13). We shall investigate both possibilities, but before we proceed we introduce some notation:

$$\hat{y} := Xb; \quad e_y := y - \hat{y} = \{I - X(X'V^{-1}X)^{-1}X'V^{-1}\} y$$

$$\hat{z} := Wb; \quad e_z := z - \hat{z} = \{I - W(W'W)^{-1}W'\} z = N_w u,$$

where $N_w := I - W(W'W)^{-1}W'$. Clearly $\hat{z}'e_z = 0$, $N_w W = 0$ and $e_z = V^{-1/2} e_y$. Let us now look at the asymptotic properties of

$$R_{\hat{y}}^2 := (\hat{y}'N_s y)^2 / (\hat{y}'N_s \hat{y} \cdot y'N_s y).$$

Lemma 4

If the conditions (2.9) through (2.11) obtain we have

(i) $\text{Plim } \dfrac{1}{n} (\hat{y}'N_s y - \beta'X'N_s X\beta) = 0$

(ii) $\text{Plim } \dfrac{1}{n} (\hat{y}'N_s \hat{y} - \beta'X'N_s X\beta) = 0$

(iii) $\text{Plim } \dfrac{1}{n} (y'N_s y - \beta'X'N_s X\beta - \sigma^2 \text{tr} V) = 0$.

If additionally we also assume

$$\lim_{n\to\infty} \frac{1}{n} X'N_s X = H \text{ and } \lim_{n\to\infty} \frac{1}{n} \text{tr } V = a, \tag{2.14}$$

then $\text{Plim } R_{\hat{y}}^2 = \beta'H\beta/(\beta'H\beta + a\sigma^2)$.
Further $\lim E R^2 = \beta'H\beta/(\beta'H\beta + a\sigma^2)$ and $D(R^2) \to 0$ as $n \to \infty$.
Hence $\text{Plim } R_{\hat{y}}^2 = 1$ iff $\sigma^2 = 0$.

Proof

$$\hat{y}'N_s y = \hat{y}'V^{-1/2}V^{1/2}N_s V^{1/2}V^{-1/2}y = \hat{z}'V^{1/2}N_s V^{1/2}z =$$

$$\beta'W'V^{1/2}N_s V^{1/2}W\beta + \beta'W'V^{1/2}N_s V^{1/2}W(W'W)^{-1}W'u + \beta'W'V^{1/2}N_s V^{1/2}u$$

$$+ u'W(W'W)^{-1}W'V^{1/2}N_s V^{1/2}u.$$

This yields

$$\frac{1}{n}(\hat{y}'N_s y - \beta'X'N_s X\beta) = \frac{1}{n} \beta'W'V^{1/2}N_s V^{1/2}W(W'W)^{-1}W'u$$

$$+ \frac{1}{n} \beta'W'V^{1/2}N_s V^{1/2}u + \frac{1}{n} u'W(W'W)^{-1}W'V^{1/2}N_s V^{1/2}u.$$

Now

$$D\{\frac{1}{n} \beta'W'V^{1/2}N_s V^{1/2}W(W'W)^{-1}W'u\} =$$

$$\frac{1}{n}\sigma^2 (\frac{1}{n}\beta'W'V^{1/2}N_s V^{1/2}W)(\frac{1}{n}W'W)^{-1}(\frac{1}{n}W'V^{1/2}N_s V^{1/2}W\beta).$$

The eigenvalues of $\frac{1}{n}W'W = \frac{1}{n}X'V^{-1}X$ are clearly bounded on the left by $\lambda_{1n}\mu_{1n}$ and by $\lambda_{kn}\mu_{nn}$ on the right. This is seen as follows. Let α_{kn} be the largest eigenvalue of $\frac{1}{n}W'W = \frac{1}{n}X'V^{-1}X$. Then

$$\alpha_{kn} = \max_p \frac{1}{n}p'X'V^{-1}Xp/p'p = \max_p \frac{1}{n}p'X'V^{-1}Xp/\frac{1}{n}p'X'Xp \cdot$$

$$\frac{1}{n}p'X'Xp/p'p \leq \max_p \frac{1}{n}p'X'V^{-1}Xp/\frac{1}{n}p'X'Xp \cdot$$

$$\max_p \frac{1}{n}p'X'Xp/p'p \leq \mu_{nn} \lambda_{kn} \leq M^2 < \infty.$$

By the same reasoning $\alpha_{1n} \geq \lambda_{1n}\mu_{1n} \geq m^2 > 0$, where α_{1n} is the smallest eigenvalue of $\frac{1}{n}W'W = \frac{1}{n}X'V^{-1}X$.

So

$$D\{\frac{1}{n}\beta'W'V^{1/2}N_s V^{1/2}W(W'W)^{-1}W'u\} \leq$$

$$\frac{\sigma^2}{m^2}(\frac{1}{n}\beta'W'V^{1/2}N_s V^{1/2})(\frac{1}{n}WW')(\frac{1}{n}V^{1/2}N_s V^{1/2}W\beta) \leq$$

$$\frac{\sigma^2 M^2}{m^2}\frac{1}{n^2}\beta'W'V^{1/2}N_s VN_s V^{1/2}W\beta \leq$$

$$\frac{\sigma^2 M^2}{m^3}\frac{1}{n^2}\beta'X'N_s X\beta \leq \frac{\sigma^2 M^2}{m^3}\frac{1}{n^2}\beta'X'X\beta \to 0$$

as $n \to \infty$ by assumption (2.9).

Further $D(\frac{1}{n}\beta'W'V^{1/2}N_s V^{1/2}u) = \frac{1}{n^2}\sigma^2\beta'W'V^{1/2}N_s VN_s V^{1/2}W\beta$

$$\leq \frac{1}{mn^2}\sigma^2\beta'W'V^{1/2}N_s V^{1/2}W\beta \leq \frac{1}{mn^2}\sigma^2 \beta'W'VW\beta = \frac{1}{mn^2}.$$
$\sigma^2\beta'X'X\beta \to 0$ as $n \to \infty$.

Now

$$\frac{1}{n} u'W(W'W)^{-1}W'V^{1/2}N_s V^{1/2}u =$$

$$(\frac{1}{n} u'W)(\frac{1}{n} W'W)^{-1}(\frac{1}{n} W'V^{1/2}N_s V^{1/2}u).$$

$$D(\frac{1}{n} u'W) = \frac{1}{n^2} \sigma^2 W'W \to 0 \text{ as } n \to \infty;$$

$$D(\frac{1}{n} W'V^{1/2}N_s V^{1/2}u) = \frac{1}{n^2} \sigma^2 W'V^{1/2}N_s VN_s V^{1/2}W =$$

$$\frac{1}{n^2} \sigma^2 X'N_s VNsX \to 0$$

by the same argument as above.

The proof of (ii) goes in the same manner. We shall now turn to $y'N_s y$. We find that

$$y'N_s y = \beta'W'V^{1/2}N_s V^{1/2}W\beta + 2 \beta'W'V^{1/2}N_s V^{1/2}u + u'V^{1/2}N_s V^{1/2}u$$

yielding

$$\frac{1}{n}(y'N_s y - \beta'X'N_s X\beta - \sigma^2 \text{ tr } V) = \frac{2}{n} \beta'W'V^{1/2}N_s V^{1/2}u +$$

$$\frac{1}{n}(u'V^{1/2}N_s V^{1/2}u - \sigma^2 \text{ tr } V).$$

Clearly Plim $\frac{1}{n} \beta'W'V^{1/2}N_s V^{1/2}u = 0$, as was proved above.

Further $u'V^{1/2}N_s V^{1/2}u = \varepsilon'N_s\varepsilon = \varepsilon'\varepsilon - (\frac{1}{n} s'\varepsilon)^2$,

$$D(\frac{1}{n} s'\varepsilon) = \frac{\sigma^2}{n^2} s'Vs \to 0 \text{ as } n \to \infty \text{ by (2.10)},$$

and

Plim $\frac{1}{n}(\varepsilon'\varepsilon - \sigma^2 \text{ tr } V) = 0$ by (2.11).

This concludes the proof of Lemma 4. Next we shall assess the asymptotic properties of

$$R_{\hat{z}}^2 := (\hat{z}'N_s z)^2 / \hat{z}'N_s \hat{z} \cdot z'N_s z. \tag{2.15}$$

Lemma 5

If assumptions (2.9) and (2.10) hold and

$$\text{Plim} \frac{1}{n} u'u = \sigma^2, \text{ then} \tag{2.16}$$

(i) $\text{Plim} \dfrac{1}{n} (\hat{z}'N_s z - \beta'W'N_s W\beta) = 0;$

(ii) $\text{Plim} \dfrac{1}{n} (\hat{z}'N_s \hat{z} - \beta'W'N_s W\beta) = 0;$

(iii) $\text{Plim} \dfrac{1}{n} (z'N_s z - \beta'W'N_s W\beta - \sigma^2) = 0.$

If in addition

$$\text{Plim} \frac{1}{n} W'N_s W = H_w, \tag{2.17}$$

then

$$\text{Plim } R_z^2 = \beta'H_w\beta / (\beta'H_w\beta + \sigma^2).$$

Proof

$$\frac{1}{n} (\hat{z}'N_s z - \beta'W'N_s W\beta) =$$

$$\frac{1}{n} \beta'W'N_s u + \frac{1}{n} \beta'W'N_s W(W'W)^{-1}W'u +$$

$$\frac{1}{n} u'W(W'W)^{-1}W'Nu.$$

As the eigenvalues of $\dfrac{1}{n} W'W$ are bounded by virtue of assumptions (2.9) and (2.10), the proof of (i) and (ii) is almost verbatim the proof of (i) and (ii) of Lemma 3. Let us then look at (iii). Here the same applies with one difference. We have to use assumption (2.16) instead of Khintchine's law to 'prove' that $\text{Plim} \dfrac{1}{n} u'u = \sigma^2$, and assumption (2.11) to prove that $\text{Plim} \dfrac{1}{n} s'u = 0.$

We will now look at an alternative goodness-of-fit measure for the heteroskedastic model

$$y = X\beta + \varepsilon; \quad E\varepsilon = 0, \quad D(\varepsilon) = \sigma^2 V.$$

Buse (1973) defines the following goodness-of-fit measure in a notation due to Maddala (1983)

$$R^2 = (WSSR_c - WSSR_u)/WSSR_c, \tag{2.18}$$

where

$$WSSR_c = y'V^{-1}y - (y'V^{-1}s)^2/s'V^{-1}s$$

and

$$WSSR_u = (y - Xb)'V^{-1}(y - Xb)$$

as defined in our notation.

Lemma 6

If assumptions (2.9), (2.10) and (2.16) hold we have

$$\text{Plim } \{\frac{WSSR_c}{n} - \frac{1}{n} WSSR_u - \frac{1}{n}\beta'X'V^{-1}X\beta$$

$$+ \frac{1}{n} \frac{(\beta'X'V^{-1}s)^2}{s'V^{-1}s}\} = 0$$

$$\text{Plim } \frac{WSSR_u}{n} = \sigma^2.$$

Obviously, if $\sigma^2 = 0$, then $R^2 = 1$.

Proof

We have

$$WSSR_c - WSSR_u = y'V^{-1}y - y'V^{-1}y + 2\ y'V^{-1}Xb -$$
$$b'X'V^{-1}Xb - (y'V^{-1}s)^2/s'V^{-1}s =$$
$$2y'V^{-1}Xb - b'X'V^{-1}Xb - (y'V^{-1}s)^2/s'V^{-1}s =$$
$$2y'V^{-1}X(X'V^{-1}X)^{-1}X'V^{-1}y - y'V^{-1}X(X'V^{-1}X)^{-1}X'V^{-1}y -$$
$$(y'V^{-1}s)^2/s'V^{-1}s =$$
$$y'V^{-1}X(X'V^{-1}X)^{-1}X'V^{-1}y - (y'V^{-1}s)^2/s'V^{-1}s =$$
$$\beta'X'V^{-1}X\beta + \varepsilon'V^{-1}X(X'V^{-1}X)^{-1}X'V^{-1}\varepsilon + 2\beta'X'V^{-1}\varepsilon -$$
$$(\beta'X'V^{-1}s + \varepsilon'V^{-1}s)^2/s'V^{-1}s.$$

Then

$$\text{Plim } \frac{1}{n}\beta'X'V^{-1}\varepsilon = 0, \text{ because}$$

$$E(\frac{1}{n}\beta'X'V^{-1}\varepsilon) = 0 \text{ and } D(\frac{1}{n}\beta'X'V^{-1}\varepsilon) =$$

$$\frac{\sigma^2}{n^2}\beta'X'V^{-1}X\beta = \frac{\sigma^2}{n}\beta'(\frac{1}{n}X'V^{-1}X)\beta \leq \frac{\sigma^2}{n}\alpha_{kn}\beta'\beta \leq$$

$$M^2 \frac{\sigma^2}{n}\beta'\beta \to 0 \text{ as } n \to \infty$$

(the boundedness of α_{kn} the greatest eigenvalue of $\frac{1}{n}X'V^{-1}X$ has been proven in lemma 3).

$\text{Plim } \frac{1}{n}\varepsilon'V^{-1}X(X'V^{-1}X)^{-1}X'V^{-1}\varepsilon = 0$, because

$$\frac{1}{n}\varepsilon'V^{-1}X(X'V^{-1}X)^{-1}X'V^{-1}\varepsilon = \frac{1}{n}u'W(W'W)^{-1}W'u$$

$$= (\frac{1}{n}u'W)(\frac{W'W}{n})^{-1}(\frac{W'u}{n}) \text{ and as we have already}$$

shown $D(\frac{1}{n}u'W) \to 0$ as $n \to \infty$ and $(\frac{W'W}{n})^{-1}$ remains finite.

The rest of the proof requires us to show that $\frac{1}{n}(\beta'X'V^{-1}s + \varepsilon'V^{-1}s)^2/s'V^{-1}s$ remains finite and that $\text{Plim } \frac{\varepsilon'V^{-1}s}{n \, s'V^{-1}s} = 0$.

Now

$$\frac{1}{n}(\beta'X'V^{-1}s + \varepsilon'V^{-1}s)^2/s'V^{-1}s =$$

$$\{\frac{1}{n}(\beta'X'V^{-1}s + \varepsilon'V^{-1}s)\}^2/\frac{s'V^{-1}s}{n}\} =$$

$$(\frac{\beta'X'V^{-1}s}{n})^2 + \frac{2\beta'X'V^{-1}s \, \varepsilon'V^{-1}s}{n} + (\frac{\varepsilon'V^{-1}s}{n})^2\}/$$

$$\frac{s'V^{-1}s}{n}.$$

Now $D(\frac{\varepsilon'V^{-1}s}{n}) = \frac{\sigma^2}{n^2}s'V^{-1}s \to 0$ as $n \to \infty$ and $\frac{s'V^{-1}s}{n}$ is bounded away from zero, so clearly we have

$\text{Plim } (y'V^{-1}s)^2/n \, (s'V^{-1}s) - (\beta'X'V^{-1}s)^2/n \, (s'V^{-1}s) = 0.$

$\text{WSSR}_u = (y - Xb)' \, V^{-1}(y - Xb) =$

$y' \, (I - X(X'V^{-1}X)^{-1}X'V^{-1})' \, V^{-1}(I - X(X'V^{-1}X)^{-1}X'V^{-1})y =$

$y' \, (V^{-1} - V^{-1}X(X'V^{-1}X)^{-1}X'V^{-1})(I - X(X'V^{-1}X)^{-1}X'V^{-1})y.$

Now
$$y'(V^{-1} - V^{-1}X(X'V^{-1}X)^{-1}X'V^{-1})y =$$
$$(X\beta + \varepsilon)'(V^{-1} - V^{-1}X(XV^{-1}X)^{-1}X'V^{-1})(X\beta + \varepsilon) =$$
$$\varepsilon'(V^{-1} - V^{-1}X(X'V^{-1}X)^{-1}X'V^{-1})\varepsilon$$

and $\text{Plim} \dfrac{1}{n}\varepsilon'(V^{-1} - V^{-1}X(X'V^{-1}X)^{-1}X'V^{-1})\varepsilon = \sigma^2$

in view of assumption (2.16) and using the fact that $\text{Plim} \dfrac{\varepsilon'V^{-1}X}{n} = 0$ and that $(\dfrac{X'V^{-1}X}{n})^{-1}$ remains finite.

3.

This Section is devoted to the conclusions we can draw from the preceding two paragraphs and also discusses the necessity of some ramifications. We have shown that under quite general conditions R^2 will be a mean-squared-error consistent estimate of some complicated function of β, X and σ^2. The value of this limit depends solely on the ratio $\sigma^2/\beta'H\beta$. So whether R^2 is near unity for large enough n is completely determined by the ratio above. For R^2 as defined by Buse we cannot say very much about its asymptotic behaviour unless we make some very drastic assumptions about the value of $(\beta'X'V^{-1}s)^2/n(s'V^{-1}s)$ as n tends to infinity. The reader may wonder why we have not looked at the asymptotic distribution (if any) of a suitably normed R^2. Having established that R^2 is not a veridical goodness-of-fit measure, this would seem a vicissitude. Another aspect that the paper neglects is that of stochastic regressors. If one would be willing to assume independence of the random regressors and the errors and further assume:

$\text{Plim} \dfrac{1}{n} X'N_sX = \lim \dfrac{1}{n} EX'N_sX = H$ and

$\text{Plim} \dfrac{1}{n} X'X = Q$ a p.d. matrix, the final result of lemma 3 would still hold.

Of course one should realize that in a model with non-random regressors, the choice of the regressors cannot influence σ^2 while in the model with stochastic regressors the choice of regressors can exert a great influence on σ^2, but a different choice of the random regressors would probably also change β and H and the value of the probability limit of R^2 is still determinated by the ratio $\sigma^2/\beta'H\beta$.

Bibliography

Barten AP. 1987. The coefficient of determination for regressions without a constant term. This volume.

Buse A. 1973. Goodness of fit in generalized least squares estimation. American Statistician, 27: 106–108.

Cramer JS. 1987. Mean and variance of R^2 in small and moderate samples. Journal of Econometrics (forthcoming).

Koerts J and Abrahamse APJ. 1970. The correlation coefficient in the general linear model. European Economic Review, 1: 410–427.

Lukacs E. 1975. Stochastic Convergence (second edition). New York: Academic Press.

Maddala GS. 1983. Limited-dependent and qualitative variables in econometrics. Cambridge: Cambridge University Press.

Modelling multivariate stochastic time series for prediction: another look at the Lydia Pinkham data

MARTIN M.G. FASE

1. Introduction

Market research very often considers different time series simultaneously and analyses their interdependence within the regression framework. The most prominent example is the relationship between sales and advertisement of which the carryover effect of advertising on sales has attracted much interest. Clarke's (1976) well-known survey lists 69 studies up to 1975 and Aaker & Carman (1982) review several additional studies since then. The main emphasis in these studies has been on proper statistical modelling of the lag structure in the spirit of the empirical-econometrics tradition. The most common statistical specification has been the Koyck model for adaptive behaviour.

A less behaviourally oriented approach is provided by the Box-Jenkins time-series framework. Examples from the sales-advertisement domain are Bretschneider, Carbone & Longini (1982), and Kapoor, Madhok & Wu (1981). These authors use the univariate time-series or transfer-function models, which however, do not allow for feedback. Nevertheless, they offer an opportunity to analyse the direction of causality. An early application of the Granger-Sims causality approach to the sales-advertising relationship is offered by e.g. Ashley, Granger & Schmalensee (1980). A straightforward generalization of the univariate stochastic time series model is provided by the multivariate stochastic time-series model. The few published market-research applications of this model [see e.g. Aaker, Carman & Jacobson (1982), Bhattacharyya (1982), Hanssens (1980), and Umashankar & Ledolter (1983)] have in common a two-step procedure for identification. The first step attempts to fit univariate ARIMA models to each of the time series studied. The second step employs the cross correlations of the estimated univariate innovations, i.e. the prewhitened series, to identify either the multivariate model or the direction of causality. In an interesting theoretical paper, Maravall (1981) has pointed out that this two-step procedure is likely to complicate the analysis and to lead to a seriously misspecified model. Therefore, Tiao & Box (1981) offer an alternative tool which avoids this cumbersome two-step procedure, but starts from non-prewhitened series.

One purpose of this article is to investigate the latter approach empirically with the aid of the well-known sales-advertising data of Lydia Pinkham and to compare the resulting models with those of other investigators employing some two-step methodology. A second aim is to investigate the influence of aggregation over time empirically in the sales-advertising framework.

The plan of this paper is as follows. Section 2 sets out briefly the statistical methodology behind the multivariate stochastic model. Section 3 presents the empirical modelling of the Lydia Pinkham sales-advertising time-series data, which are described briefly also. In section 4 our estimation results are compared with previous results of others. An important motivation for our research was the desire to evaluate the predictive performance of the multivariate model vis-à-vis other approaches. Our conclusion is that for the Lydia Pinkham data the multivariate model does not improve forecasting ability substantially, but shows strong feedback among monthly sales and advertising. Thus, this paper is mainly an empirical application of the multivariate time series model to micro-economic data, which, in contrast to applications to macro-economic and monetary data [see e.g. Guilkey and Salemi (1982); Hsiao (1979)], seems to be rare.

2. Statistical methodology

2.1. The stochastic multivariate time series model

The behaviour of a univariate time series x_t is often described by stochastic difference equations or ARIMA (p, d, q) models of the form:

$$\varphi_p(B)\nabla^d x_t = \theta_o + \theta_q(B)a_t \qquad (1)$$

where $\varphi_p(B) = 1 - \varphi_1 B - \varphi_2 B^2 - \ldots - \varphi_p B^p$, $\theta_q(B) = 1 - \theta_1 B - \theta_2 B^2 \ldots - \theta_q B^q$ are scalar polynomials in B, θ_o is a constant. B the backshift operator such that $Bx_t = x_{t-1}$, ∇ the difference operator such that $\nabla = 1 - B$, and p, d, q are nonnegative integers. In (1) the a_t are stochastically independent, normally distributed random shocks with zero mean and constant variance σ_a^2.

Apart from differencing the straightforward generalization of (1) is the $m \times 1$ vector or multivariate stochastic difference equation:

$$\varphi(B)x_t = \theta_o + \theta(B)a_t \qquad (1a)$$

where $\varphi(B) = I - \varphi_1 B - \varphi_2 B^2 - \ldots - \varphi_p B^p$, $\theta(B) = I - \theta_1 B - \theta_2 B^2 - \ldots - \theta_q B^q$ are matrix polynomials in the backshift operator B, the φ and θ matrices of order $m \times m$, I the $m \times m$ identity matrix, θ_{oo} a $m \times 1$ vector of constants,

$x_t = (x_{1t}, x_{2t}, \ldots, x_{mt})'$ a $m \times 1$ vector of observations at time t on a set of m univariate jointly stationary time series, $a_t = (a_{1t}, a_{2t}, \ldots, a_{mt})'$, a $m \times 1$ vector of stochastically independent and normally distributed random shocks with zero mean and covariance matrix $\sum_a$, i.e. $Ea_t = 0$ and $E(a_t a_t') = \sum_a$, where $\sum_a$ is a positive definite matrix of order $m \times m$. Equation (1a) is called a vector ARMA model. It is well-known that the parameters of the vector ARMA model must meet certain conditions to ensure stationarity and invertibility, i.e. the roots of the determinantal equation of both $|\varphi(B)| = 0$ and $|\theta(B)| = 0$ should all lie outside the unit circle.

An alternative but equivalent representation of (1a) is:

$$\begin{pmatrix} \varphi_{11}(B) & \ldots & \varphi_{1m}(B) \\ \cdot & & \cdot \\ \cdot & & \cdot \\ \varphi_{m1}(B) & \ldots & \varphi_{mm}(B) \end{pmatrix} \begin{pmatrix} x_{1t} \\ \cdot \\ \cdot \\ x_{mt} \end{pmatrix} = \begin{pmatrix} \theta_1 \\ \cdot \\ \cdot \\ \theta_m \end{pmatrix} + \begin{pmatrix} \theta_{11}(B) & \ldots & \theta_{1m}(B) \\ \cdot & & \cdot \\ \cdot & & \cdot \\ \theta_{m1}(B) & \ldots & \theta_{mm}(B) \end{pmatrix} \begin{pmatrix} a_{1t} \\ \cdot \\ \cdot \\ a_{mt} \end{pmatrix} \quad (1b)$$

with $\varphi_{ij}(B)$ and $\theta_{ij}(B)$ scalar polynomials in B of degree p_{ij} and q_{ij} respectively.

The polynomials in the diagonal position start with unity, while the polynomials in the off-diagonal position start with terms which are some power of B.

To illustrate the above notation and for future reference we consider here three important special cases of the vector ARMA model, which very often occur in practice, viz. the bivariate AR model, the bivariate MA model and the bivariate ARMA model. The bivariate AR model reads:

$$\begin{pmatrix} \varphi_{11}(B) & \varphi_{12}(B) \\ \varphi_{21}(B) & \varphi_{22}(B) \end{pmatrix} \begin{pmatrix} x_{1t} \\ x_{2t} \end{pmatrix} = \begin{pmatrix} a_{1t} \\ a_{2t} \end{pmatrix}$$

with $\varphi_{ij}(B) = \sum_{k=0}^{p_{ij}} -\varphi_{ij,k} B^k$, where, as mentioned before, for $k = 0$, $-\varphi_{ij,k} = 1$ for $i = j$ and where $\varphi_{ij,k} = 0$ otherwise, with p_{ij} the degree of the polynomial element i, j. Thus a bivariate AR(1) model, deleting the suffix k for the sake of simplicity here, is:

$$\begin{pmatrix} (1 - \varphi_{11}B) & -\varphi_{12}B \\ -\varphi_{21}B & (1 - \varphi_{22}B) \end{pmatrix} \begin{pmatrix} x_{1t} \\ x_{2t} \end{pmatrix} = \begin{pmatrix} a_{1t} \\ a_{2t} \end{pmatrix}$$

However, it should be noted that the p_{ij} are not necessarily equal to each other for each i and j as is here the case.

The second model often encountered in practice is the bivariate moving-average or MA(2) model.

$$\begin{pmatrix} x_{1t} \\ x_{2t} \end{pmatrix} = \begin{pmatrix} (1 - \theta_{11,1}B - \theta_{11,2}B^2) & (-\theta_{12,1}B - \theta_{12,2}B^2) \\ (-\theta_{21,1}B - \theta_{21,2}B^2) & (1 - \theta_{22,1}B - \theta_{22,2}B^2) \end{pmatrix} \begin{pmatrix} a_{1t} \\ a_{2t} \end{pmatrix}$$

Similarly and again deleting the third suffix to keep the notation simple, a bivariate ARMA(1, 1) model reads as:

$$\begin{pmatrix} 1 - \varphi_{11}B & -\varphi_{12}B \\ -\varphi_{21}B & 1 - \varphi_{22}B \end{pmatrix} \begin{pmatrix} x_{1t} \\ x_{2t} \end{pmatrix} = \begin{pmatrix} 1 - \theta_{11}B & -\theta_{12}B \\ -\theta_{21}B & 1 - \theta_{22}B \end{pmatrix} \begin{pmatrix} a_1 \\ a_2 \end{pmatrix} \quad (1c)$$

2.2 Conditions for causality

Since Sims' (1972) seminal article on the causality between money and income the time series framework has become an important statistical tool for empirical causality analysis in economics along the lines suggested by Granger (1969).

Of course, this does not mean that there is full agreement among economists about the appropriateness of Granger's definition of causality [see Zellner (1979)]. The multivariate time series model offers a natural generalization for this purpose. Following Kang (1981), Hsiao (1982) and Osborn (1984), necessary and sufficient conditions for causality in the multivariate ARMA process can be formulated. These are formalizations of the intuitive notion that in a multivariate framework causality means something like an upper-block-triangular structure of the matrices φ(B) and θ(B) in (1b). Suppose that in this general ARMA model x and a can be partitioned so that x_1 and a_1 are g × 1 vectors and x_2 and a_2 are h × 1 vectors, g + h = m, with submatrices $\Phi(B)_{ij}$ and $\theta(B)_{ij}$ conformably partitioned. Then, according to Osborn, the variables in x_1 do not cause the variables in x_2 if and only if there is a representation of the multivariate ARMA system (1b) in which the submatrices $\varphi(B)_{21} = \theta(B)_{21} = 0$. In other words, x_1 does not cause x_2 if an only if in x = φ(B)$^{-1}$θ(B)a either the polynomial matrix φ(B)$^{-1}$θ(B) is upper-block-triangular when partitioned conformably to x and a or, equivalently, in a = θ(B)$^{-1}$φ(B) x the polynomial matrix θ(B)$^{-1}$φ(B) is upper-block-triangular, when partitioned conformably.

Using the bivariate ARMA model (1c) as an example to illustrate the above, the necessary and sufficient condition for x_1 not to cause x_2 is:

$$-\theta_{21} + \varphi_{11}\theta_{21} - \varphi_{21}\theta_{11} + \varphi_{21} = 0 \quad (2)$$

with φ_{ij} and θ_{ji} elements of the coefficient matrices in (1c). We note that this condition is weaker than the condition $\varphi_{21} = \theta_{21} = 0$ which is of course sufficient, but, in spite of its strong intuitive appeal, not necessary.

2.3 Specification and estimation

The class of vector ARMA models contains a large number of possible models. Therefore, our aim is to specify a parsimonious model representing the data as adequately as possible. Parsimony here means a model with as few parameters as possible. The main instrument of model selection, i.e. choosing numerical values of the integers p and q of the polynomials in the backshift operator B, includes the sample cross-correlation and partial-autocorrelation matrices, which have to be estimated from the data. A useful device for determining the order of the model is the pattern of the indicator symbols of the sample cross-correlations and of the partial-autoregressive matrices and related statistics suggested by Tiao & Box (1981) and technically simplified for practical use in Tiao & Tsay (1983).

As soon as a suitable tentative specification has been chosen, efficient estimates of the parameter matrices φ, θ and $\sum_a$ are obtained by maximizing the exact likelihood function L with

$$L(\varphi, \theta, \sum_a | x) \propto L_c(\varphi, \theta, \sum_a | x) L_1(\varphi, \theta, \sum | x)$$

where

$$L_c(\varphi, \theta, \sum_a | x) \propto |\sum_a|^{-(n-p)/2} \exp\{-1/2 \, \text{tr} \, \sum_a^{-1} S(\varphi, \theta)\}$$

is the conditional likelihood function with $S(\varphi, \theta) = \sum_{t=p+1}^{n} a_t a_t'$, the a_t following from the vector ARMA(p, q) model to be estimated and L_1 depending only on the data vectors $x_1, \ldots, x_p$ if all $q = 0$ or on all data vectors $x_1, \ldots, x_n$ if $q \neq 0$, with n the number of vector observations. Because estimation by using the exact likelihood function is rather slow, the conditional likelihood function is used in the preliminary stage of model building and the exact method mainly towards the end [see also Tiao & Box (1981, p. 809)].

3. The empirical analysis

We have analyzed two series of both monthly and yearly data. These two series are advertising expenditure, x_{1t}, by the Lydia Pinkham Medicine Company and dollar sales, x_{2t}, of the Lydia Pinkham Medicine Company. These data see to have a similar reputation among market researchers as the famous hogcorn data have among economists or mink-muskrat data among ecologists, mainly because long series have been collected [see Nerlove *et al.* (1979), Jenkins (1974) and Chan & Wallis (1978)]. Because of the possible unfamiliarity with the Lydia Pinkham data we start with a brief discussion of these data.

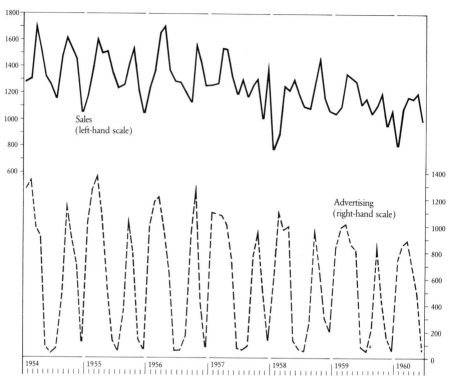

Fig. 1. Domestic U.S. sales and advertising expenditures by Lydia Pinkham Medicine Company 1907–1960 (in $1000).

3.1 The Lydia Pinkham data

The Lydia Pinkham data refer to a medicinal remedy against menopausal malaise and menstrual pain, which has been on the market since 1873. The data base, described in Palda (1964), became public because of controversies over the product and a major court case. This publicly available data base has on many occasions been used by marketing researchers for several reasons. First, the time series is sufficiently long both for the yearly and the monthly data. Second, for the Lydia Pinkham Medicine Company advertising was the exclusive marketing instrument while the absence of direct competitors at the time for this product makes the market a closed sales-advertising system. Of course, this means that advertising in this was mainly to inform the public i.e. to create a market for the product. Third, the product is a low-cost consumer product, which is frequently purchased while price changes are small and rare. Thus analysis is not obscured by any inherent trend, while the famous Marshallian *ceteris paribus* condition seems to be fulfilled.

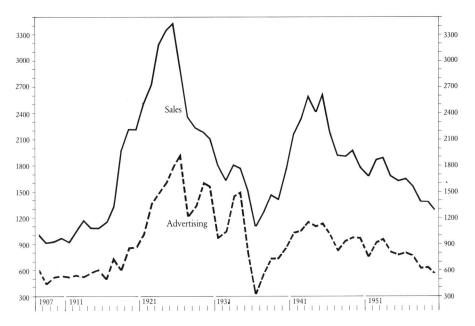

Fig. 2. Domestic U.S. monthly sales and advertising expenditures by Lydia Pinkham Medicine Company 1954–1960 (seasonally unadjusted, in $1000).

The yearly data refer to the sample period 1907–1960; the monthly data to 1954(1)–1960(6). In the estimation we used the yearly data for 1907–1954 and the montly data for January 1954 to December 1958. Thus we are left with six yearly and eighteen monthly observations to examine the *ex post* forecasting performance of the estimated models. Previous studies mainly analysed the posssibility to predict sales as response to advertisement expenditure within a regression framework. Here we study the structure by jointly modelling the two series. Using the Palda (1964)-series we have depicted both the yearly and the monthly series over the whole sample periods in Figs 1 and 2, in order to give some feeling for this data set.

Sofar, the Lydia Pinkham data are a unique publicly available source of information on commercial data which have been exploited by researchers very often [see e.g. Bhattacharyya (1982), Caines, Sethi & Brotherton (1977), Hanssens (1980), Helmer & Johansson (1977)]. Hopefully, our analysis may encourage others to uncover similar data for statistical analysis on a micro level, enabling time-series analysts to implement the theoretical time-series model to a variety of meaningful practical cases and to examine the implications of aggregation over time empirically [see also Abraham (1982), Brewer (1973) and Wei (1978)].

Table 1. Pattern of sample cross correlations for Lydia Pinkham data over 12 lags.

	x_{1t}	x_{2t}
x_{1t}	+++++. . . . . . −	+++. . . . −−−−−
x_{2t}	++++++. . . . . −	++++. . . +−−−−

Note: the + and − mean significant positive and negative cross correlations respectively; the . indicates insignificant cross correlations at the 5% significance level.

3.2 Analysis of yearly data

Table 1 shows the pattern of sample cross correlations for the yearly Lydia Pinkham data, using the device of indicator symbols of Tiao & Box (1981) as a tool of identification.

The persistence of correlations shown in Table 1 indicates autoregressive behaviour. To approximate the order l of the autoregressive model tentatively, the likelihood-ratio statistic, $M(l)$, is employed, testing the null hypothesis $\varphi_l = 0$ against $\varphi_l \neq 0$ when AR models of increasing order l are fitted. Table 2 shows this statistic together with the sample partial-autoregression matrices, $\hat{\varrho}(l)$, (indicated by symbols) and the diagonal elements of the sample residual-covariance matrices $\hat{\Sigma}_a$.

Table 2 indicates that a vector AR(1) model would be adequate, but we also

Table 2. Pattern of partial-autoregression matrices and the other statistics.

l	$\hat{\varrho}(l)$	$M(l)$	residual variances × 10^3
1	+ −	88.05	5.29
	+ +		4.12
2	. .	8.51	4.17
	. .		3.92
3	. .	14.01	4.14
	+ .		2.72
4	. .	3.21	4.05
	. .		2.62
5	. .	3.14	3.70
	. .		2.45
6	. .	1.70	3.54
	. .		2.35

Note: $M(l)$ follows a χ^2 distribution with 4 degrees of freedom. 5% critical value is 9.5; 1% critical value is 13.3 In previous notation l is a scalar with $l = \max(p_{ij})$.

Table 3. Pattern of residual cross-correlations of fitted vector AR(1) model over 24 lags.

	a_1	a_2
a_1	. . . .	. . . .
a_2	+. . . . . +. . −+. . .	. . . . −. . . . . .

fitted a vector AR(3) model. The pattern of residuals of the latter model shows white noise, but this is not the case for the fitted AR(1) model as is brought out by Table 3.

In view of this result we decided also to proceed with a multivariate ARMA(1, 1) model as an alternative to the suggested vector AR(3) model. A summary of our results is given in Table 4 where we preferred the conditional

Table 4. Estimation results for the vector ARMA(1, 1) and the vector AR(3) model.

	ARMA(1, 1) exact likelihood						
	$\hat{\varphi}$		$\hat{\theta}_0$	$\hat{\theta}$		$\hat{\Sigma}_a \times 10^4$	
Full model	.17	.39	79.30	−.40	−.10	3.12	
	(.23)	(.13)	(113.3)	(.22)	(.15)		
	−.76	1.26	260.50	−.59	−.10	1.34	3.84
	(.27)	(.15)	(122.8)	(.28)	(.20)		
Restricted	0	.48	71.90	−.51	0	3.15	
final model		(.06)	(114.5)	(.12)			
	−.92	1.34	254.5	−.73	0	1.36	3.85
	(.18)	(.10)	(120.7)	(.22)			

	AR(3) conditional likelihood								
	$\hat{\varphi}_1$		$\hat{\varphi}_2$		$\hat{\varphi}_3$		$\hat{\theta}_0$		
Full model	.51	.58	−.35	−.55	.26	.30	−37.5	2.51	
	(.14)	(.14)	(.16)	(.20)	(.15)	(.14)	(81.0)		
	−.17	1.34	−.29	−.30	.07	.04	255.9	1.49	3.88
	(.17)	(.17)	(.20)	(.25)	(.18)	(.18)	(101.1)		
Restricted	.57	.44	−.25	−.46	.24	.29	−56.9	2.72	
final model	(.12)	(.12)	(.14)	(.18)	(.13)	(.13)	(82.5)		
	0	.90	0	0	0	0	208.7	2.14	5.82
		(.06)					(113.2)		

Note: restricted here means setting the insignificant estimates in the unrestricted or full model at zero.

likelihood over the exact likelihood in the AR(3)-model for computational reasons.

It is easily verified that the roots of the determinantal equations of both the ARMA(1, 1) model and the AR(3) model lie all outside the unit circle ensuring stationarity and invertibility of the estimated models.[1] Diagnostic checking of the residuals shows that, besides cross correlation, the residuals of the ARMA(1, 1) model are white noise, and these of the AR(3) model almost white noise. Therefore, we also estimated a vector ARMA(2, 1) model but this led to this model being rejected because of overparameterizing and the residual patterns.

The two final estimated models set out in Table 4 imply that the yearly Lydia Pinkham data are described by the following bivariate stochastic processes:

$$x_{1t} \quad - .48x_{2t-1} = 71.9 + a_{1t} + .51a_{1t-1} \tag{3a}$$

$$+ .92x_{1t-1} + x_{2t} - 1.34x_{2t-1} = 254.5 \quad + .73a_{1t-1} + a_{2t} \tag{3b}$$

and

$$(1 - .57B + .25B^2 - .24B^3)x_{1t} +$$
$$+ (- .44B + .46B^2 - .29B^3)x_{2t} = -56.9 + a_{1t} \tag{3'a}$$

$$(1 - .90B)x_{2t} = 208.7 + a_{2t} \tag{3'b}$$

Both systems (3) and (3′) allow an interesting interpretation. Equation (3a) – which is a very simple transferfunction – shows advertising expenditure as lagging behind sales by one year. This means that in a univariate framework sales are causing advertisement in the Granger sense. This is an unexpected but meaningful result suggesting that sales are a leading indicator for advertising expenditure. However, as the employed framework is bivariate, this *prima facie* conclusion does not seem to be fully warranted. Application of criterion (2), even supports the contrary conclusion of advertising (x_1) causing sales (x_2). Upon substitution of (3a) into (3b) we obtain a univariate autoregressive model for sales.

System (3′) is a recursive system with error terms which are correlated. In this case sales may be rewritten as a univariate autoregressive model. However, advertising expenditure is now an univariate infinite MA process. System (3′) is a bivariate AR model with $\varphi_{12}(B) = 0$ and $\theta_{12}(B) = 0$. Thus, $\theta^{-1}\varphi$ is upper-block-triangular which implies that in this model and according to our terminology x_1 (advertising expenditure) does not cause x_2 (sales).

As to both parsimony and plausibility, model (3′) seems to score better than model (3), but this is difficult to assess on the basis of these results. A possible

---

[1] For the ARMA(1, 1) model the determinantal equation is $1 - 1.34B + .442B^2 = 0$ with roots $|w_1| = 1.71$ and $|w_2| = 1.32$. For the AR(3) model this equation is $1 - 1.47B + 0.76B^2 - 0.47B^3 + 0.24B^4 = 0$ with roots $|w_1| = |w_2| = 1.70$ and $|w_3| = |w_4| = 1.44$.

implication of both model (3) and model (3') may be that the carry-over effects of advertising within the year are small and are thus washed away with annual aggregation. To explore this a little further an analysis of monthly data is necessary.

3.3 Analysis of monthly data

The pattern of cross correlations among the monthly advertising and dollar sales data for January 1954 to December 1958, as well as the likelihood-ratio statistic $M(l)$ which – for the sake of brevity – are not reported here, suggest a bivariate seasonal vector AR model of the form:

$$(I - \varphi B)(I - \Phi B^{12}) x_t = \theta + a_t. \quad (4)$$

Estimation of system (4) leads to the initial empirical model obtained by the conditional maximum-likelihood method set out in the upper part of Table 5. Inspection of the estimated pattern of residual cross-correlations – not reported here – shows that the residuals are white noise.

In view of the estimation results we have re-estimated the unrestricted model (4) setting the insignificant estimates of φ_{ij} and Φ_{ij} at zero. The resulting restricted model, again obtained by conditional maximum-likelihood estimation, and the estimate of the variance–covariance matrix Σ_a – are shown in the lower part of Table 5.

The renewed unreported pattern of residual suggests that this final model is statistically adequate. Moreover, in this case it is easily verified numerically that the roots of the determinantal equations[2] meet the conditions of stationarity and invertibility.

Table 5. Estimation results for the vector $AR(1) \times (1)_{12}$-model.

	$\hat{\varphi}$		$\hat{\Phi}$		$\hat{\theta}_0$	$\hat{\Sigma}_a \times 10^4$	
Initial full model	.31	−.30	.71	.28	67.7	3.83	
	(.15)	(.19)	(.08)	(.19)	(251.5)		
	.36	.14	−.00	.42	712.7	0.80	1.98
	(.11)	(.14)	(.06)	(.15)	(199.0)		
Restricted final	.33	−.34	.76	0	337.2	4.16	
model	(.15)	(.19)	(.06)		(164.0)		
	.39	0	0	.41	694.0	0.86	1.97
	(.10)			(.13)	(166.7)		

Note: restricted here means setting the insignificant estimates in the unrestricted or full model at zero.

[2] These equations are $(1 - 0.33B + 0.13B^2) = 0$ and $(1 - 0.76B)(1 - 0.41B) = 0$ with roots $|w_1| = |w_2| = 2.75$ and $|w_1| = 2.44, |w_2| = 1.32$ respectively.

As an alternative to model (4) we also estimated the model

$$(I - \varphi B)(I - IB^{12}) x_t = \theta_0 + a_t \tag{5}$$

Although the estimation results of this vector ARMA model imply a residual pattern showing white noise, we rejected model (5) as a useful simplification, mainly because most coefficients are insignificant and the seasonal differences do not appear in the preliminary analysis of the data.

The final estimation of model (4) can be written as:

$$\begin{aligned}(1 - .33B)(1 - .76B^{12})x_{1t} + .34B(1 - .41B^{12})x_{2t} &= 337.2 + a_{1t} \\ .39B\,(1 - .76B^{12})x_{1t} + \quad (1 - .41B^{12})x_{2t} &= 694.0 + a_{2t}\end{aligned} \tag{6}$$

This representation implies full interaction or two-way causality between dollar sales and advertising expenditure. In other words, the simple causal structure obtained for the yearly data disappears completely for monthly series. Contrasting this result with our findings for yearly data shows that our *prima facie* conclusion of sales leading advertising expenditure certainly does not hold for monthly data. The form of (6) also shows important carry-over effects for monthly figures, supporting our conjecture that temporal aggregation over the year washes out these effects and may lead to biased results. However, what is most important is that for the Lydia Pinkham case the estimated monthly model demonstrates an unmistakable interaction between advertising and sales, which precludes the straightforward one-directional causality hypothesis underlying other studies of the Lydia Pinkham data.

3.4 The linkage between the yearly and the monthly models

The question may arise whether the yearly model follows logically from the model identified for monthly data. This is mainly a theoretical issue pursued here primarily for methodological reasons by way of an analytical excursion. There is, as indicated above, also an empirical reason for raising this question. The monthly results imply a two-way causality, which seems to contradict both the results with the yearly data and the fact that in practice advertising decisions most often are made yearly rather than monthly.

For monthly data our empirical analysis has suggested a bivariate AR-model of the form:

$$(I - \varphi B)(I - \Phi B^{12})x_t = \theta_0 + b_t \tag{7}$$

with b_t a random vector in the monthly model.

For yearly data the bivariate model obtained is of the ARMA(1, 1) form:

$$(I - \varphi B) y_T = \theta_0 + (I - \theta B) a_T \tag{8}$$

with $y_T = (y_{1T}, y_{2T})'$ the 12-month nonoverlapping aggregate of $x_t = (x_{1t}, x_{2t})$ and a_T the corresponding vector with random elements. The basic problem here is whether models (7) and (8) can be reconciled logically. The answer is positive and the reasoning is as follows:

First, straightforward multiplication shows:

$$(I - \varphi B)(I + \varphi B + \varphi^2 B^2 + \ldots + \varphi^{11} B^{11}) = (I - \varphi^{12} B^{12}).$$

Multiplying (7) with $(I + \varphi B + \ldots + \varphi^{11} B^{11})$ leads to:

$$(I - \varphi^{12} B^{12})(I - \Phi B^{12})x_t = \delta_0 + (I + \varphi B + \ldots + \varphi^{11} B^{11}) b_t \tag{9}$$

where the righthand side is an MA(0) for yearly data.

Second, by definition, the following holds:

$$y_{iT} = (1 + B + B^2 + \ldots + B^{11})x_{it} \text{ for } i = 1, 2 \tag{10}$$

Combining (9) and (10) results in:

$$(I - \varphi^{12} B^{12})(I - \Phi B^{12})y_T =$$
$$\delta_0'' + (1 + B + \ldots + B^{11})(I + \varphi B + \ldots + \varphi^{11} B^{11}) b_t \tag{11}$$

which may be interpreted approximately as a vector ARMA (1, 1) model in B^{12}, assuming the elements of $(-\varphi^{12} B^{12} + \varphi^{12} B^{24})$ to be empirically negligible.

This theoretical result is reassuring, particularly as it shows that our results for yearly data are in accordance with our monthly results and do not contradict each other.

4. Comparison with previous studies

It is instructive to compare the results of this study with the estimation results obtained by previous researchers of the time series model, who also employed the Lydia Pinkham data.

For yearly data Helmer & Johansson (1977) have fitted transfer functions to the first 40 years of data, i.e. 1907–1948, and used the years 1949–1960 to examine the forecasting ability of their models. Their best fitted models are (in our notation):

$$(1 - B)x_{2t} = \frac{.52(1 - B)}{(1 - .36B)} x_{1t} + \frac{a_{2t}}{(1 - .26B)} \tag{12a}$$

$$(1 - B)x_{2t} = (.48 + .19B)(1 - B)x_{1t} + \frac{a_{2t}}{(1 - .28B)} \tag{12b}$$

with a univariate IMA (1, 1) model for sales of the form:

$$(1 - B)x_{2t} = 21.03 + (1 + .45B)a_{2t}. \tag{12c}$$

It is obvious that only our restricted vector AR(3) model, given by equations (3'a, b), implies results comparable with Helmer & Johansson's. The univariate model of equation (12c) resembles our equation (3'b), the latter being less restrictive than Helmer & Johansson's. Equation (3'a) can be factorized and rewritten to:

$$x_{2t} = 210.7 + \frac{(1-.70B)(1+.13B+.34B^2)}{.44B(1-1.05B+.66B^2)} x_{1t} + $$

$$- \frac{1}{.44B(1-1.05B+.66B^2)} a_{1t}. \qquad (13)$$

Thus our multivariate analysis of yearly data confirms Helmer & Johansson's assumption of no feedback between sales and advertising. However, our implied transfer model (13) does not resemble their models (12a, b). We also compared the forecasting ability of our and Helmer & Johansson's equations. We found for sales a mean square error of 18.779 in ours against 16.080 or less in theirs. Thus for yearly data the multivariate modelling has not resulted in a gain in forecasting ability.

For the *monthly* Lydia Pinkham series bivariate models have been estimated earlier by Hanssens (1980) and Bhattacharyya (1982). Unlike this study, both Hanssens and Bhattacharyya carry out the two-step procedure of fitting univariate ARMA models. Hanssens, using the sample period 1954–1958, obtained the following bivariate representation (in our notation):

$$(1-B^{12})x_{1t} - .42B^{12}(1-B^{12})x_{2t} = (1-.57B^{12})a_{1t}.$$
$$-.37B(1-B^{12})x_{1t} + (1-.55B)(1-B^{12})x_{2t} = $$
$$(1-.55B)(1+.08B-.07B^2+.37B^3)a_{2t}. \qquad (14)$$

Bhattacharyya examined these results critically and concluded, contrary to Hanssens, that no effective bivariate feedback relationship could be established between the Lydia Pinkham seasonally differenced monthly dollar sales and advertising. The reason for this is that Hanssens created a unit root problem by differencing his sales series. To avoid this Bhattacharyya re-estimated Hanssens' model and included, following Palda (1964), dummies for outliers in sales and advertising. His estimated bivariate model, omitting for briefness the dummies here, is:

$$(1-.88B^{12})x_{1t} - .16B^{12}x_{2t} = -176.24 + (1+.04B-.28B^2)a_{1t}.$$
$$-.23B \quad x_{1t} + \quad x_{2t} = 1179.18 + (1+.12B-.12B^2) a_{2t}. \qquad (15)$$

Comparing systems (14) and (15) with our system (6), we may infer that both systems have interaction among sales and advertisement in common. However, we have found an error structure which is much more simple than that of Hanssens and Bhattacharyya. This implies an improvement of the *ex ante* forecasting accuracy of our model. Finally we note that, contrary to Bhattacharyya's, our model exhibits seasonal polynomials for all terms in the equation system. Thus our approach eliminates the implausible result of absent seasonality in the second equation of Bhattacharyya's system (15), which must be due to the two-step procedure, and generated the complicated error structure noted before. In all fairness it must be said that the calculated

Table 6. Inequality coefficients and mean squared error of prediction of the Lydia Pinkham series according to the models considered.

		U	U^M	U^S	U^C	MSE
Yearly models						
AR(3)	x_1	.0697	.6241	.0203	.3556	10859
	x_2	.0442	.7554	.0462	.1984	18779
ARMA(2, 1)	x_1	.0652	.7535	.0644	.1821	9518
	x_2	.0432	.7458	.0470	.2072	17963
ARMA(1, 1)	x_1	.0647	.5954	.0646	.3400	9247
	x_2	.0463	.6735	.0550	.2715	20548
H & J						
univariate	x_2	.0397	.2754	.0288	.6958	18178
bivariate	x_2	.0326	.0230	.0350	.9420	8912
Monthly models						
AR(1) i.e.	x_1	.1417	.0088	.0599	.9313	51896
model (4)	x_2	.0569	.3884	.0270	.5846	17546
AR(1) i.e.	x_1	.1638	.0041	.0072	.9887	49565
model (5)	x_2	.0621	.0262	.1283	.8455	20579
H's model						
univariate	x_1	.3148	.5631	.2213	.2156	108395
	x_2	.0640	.2135	.2095	.5770	22809
bivariate	x_1	.1385	.0584	.0228	.9188	27221
	x_2	.0671	.0002	.1428	.8570	19698
B's model						
univariate	x_1	.1118	.3500	.0000	.6500	25155
	x_2	.0690	.5038	.0750	.4212	27351
bivariate	x_1	.0949	.0901	.0182	.8918	16721
	x_2	.0828	.7818	.0417	.1766	41345

Explanatory note: $U = \dfrac{\left(\dfrac{1}{n}\sum(\hat{z}_i - z_i)^2\right)^{1/2}}{\left(\dfrac{1}{n}\sum\hat{z}_i^2\right)^{1/2} + \left(\dfrac{1}{n}\sum z_i^2\right)^{1/2}}$, with $\hat{z}_i$ predicted value and z_i actual value. Denoting $\left(\dfrac{1}{n}\sum\hat{z}_i^2\right)^{1/2} + \left(\dfrac{1}{n}\sum z_i^2\right)^{1/2} = D$

we define $U_m = \dfrac{\hat{\bar{z}} - \bar{z}}{D}$; $U_S = \dfrac{s_{\hat{z}} - s_z}{D}$;

$U_C = \dfrac{(2(1-r_{\hat{z}z})\,s_{\hat{z}}s_z)^{1/2}}{D}$ and $U^i = \dfrac{U_i^2}{U^2}$ with $i = M, S, C$.

Consequently $U^M + U^S + U^C = 1$.

220

ex post forecasting errors of Bhattacharyya's model are smaller than ours. Graphical inspection, however, shows that this is only so because of his adjustment for outliers in the sample period.

It is interesting to analyse the forecasting accuracy further by considering Theil's (1958) inequality statistic, U, rather than the mean squared error. As known, Theil's inequality coefficient allows to distinguish the proportions of inequality due to three different sources, baptised the bias U^M, the variance U^S and the covariance U^C proportions. Table 6 sets out the results of this decomposition for each of the estimated models as well as for the models of Helmer & Johansson (H&J), Hanssens (H) and Bhattacharyya (B) considered above.

For the yearly models Helmer & Johansson's scores best in terms of the overall and partial inequality coeficients, with the U^M close to zero, which is the most desirable property. However, for the monthly models our models seem to score better than those obtained from the literature, judged from the U^M which very often is close to zero with the covariance proportion U^C close to .90 in most cases.

5. Concluding remarks

In his textbook Cramer (1969, p. 4) reminds us of the fact that '... the construction of econometric models is an art. It requires an understanding of what is relevant to the particulier observations at hand far more than a wide knowledge of economic theory'. Talking on the level of aggregation and observations of particulier industries or firms he continues that 'It is a pity that data of this kind are relatively rare'. I like to connect this opinion with another *dictum* of Cramer (1978, p. 2) that '... predictions that are prompted by practical or commercial considerations ... are used as a substitute for information when decisions have to be taken in uncertain situations'. The present contribution illustrates that time-series modelling is an art indeed and that the difficulty to find data on the firm level is still present. The Lydia Pinkham data are a rare example, which however, is informative in many respects.

One may ask whether our results justify the use of the bivariate model. In view of the forecasting ability of our results with the Lydia Pinkham data, the answer must be negative, particularly as the transfer model seems to outperform the bivariate model. Again we may quote Cramer (1978, p. 5): 'If anyone can do better, let him try'. However, as far as the structure is concerned our approach has improved the empirical knowledge of the interaction between advertisement and sales. Moreover, our analysis shows that this insight is not independent of the adopted identification and estimation procedure, which is, incidentally, simpler than the two-step variety adopted by the cited and other researchers of the Lydia Pinkham data. The same holds for temporal aggregation of the data which both affects the estimation of the models and the forecasting accuracy. This leads to the basic question whether the application of the multivariate time-series model is worth the effort. The answer to this general question which is relevant to both academics and practitioners depends on what the research has in mind to attain. If one has in mind, as in this case – and in many monetary applications reviewed in Fase (1981) – the discovery of fundamental relationships between data, the answer is affirmative. However, as a case study in marketing, the effort is dubious, particularly because very often in marketing research, theory may dictate the model. In that case the time-series model is probably a too mechanistic research device. Nevertheless we believe – as we did many years ago [see Fase (1969, 1971, 1972)] – that a careful statistical model without much economic theory to start with may be a promising research strategy.

6. Acknowledgement

Helpful comments by F.J.J.S. van de Gevel and F.C. Palm are gratefully acknowledged.

Bibliography

Aaker DA and Carman JM. 1982. Are you overadvertising? *Journal of Advertising Research*, 22: 57–69.

Aaker DA, Carman JM and Jacobson R. 1982. Modelling advertising-sales relationship involving feedback: a time series analysis of six cereal brands. *Journal of Marketing Research*, 19: 116–125.

Abraham B. 1982. Temporal aggregation and time series. *International Statistical Review*, 50: 285–291.

Ashley R, Granger CWJ and Schmalensee R. 1980. Advertising and aggregate consumption: an analysis and causality. *Econometrica*, 48, 5:1149–1167.

Bhattacharyya MN. 1982. Lydia Pinkham data remodelled. *Journal of time series analysis*, 3, 2: 81–102.

Bretschneider SI, Carbone R and Longini RL. 1982. An adaptive multivariate approach to time series forecasting. *Decision Sciences*, 13, 4: 668–680.

Brewer KWR. 1973. Some consequences of temporal aggregation and systematic sampling for ARMA and ARMAX models. *Journal of Econometrics*, 1:133–154.

Caines PE, Sethi SP and Brotherton TW. 1977. Impulse response identification and causality detection for the Lydia Pinkham data. *Annals of Economic and Social Measurement*, 6: 147–163.

Chan WY and Wallis KF. 1978. Multiple time series modelling, another look at the mink-muskrat interaction. *Applied Statistics*, 27, 2: 168–175.

Clarke DG. 1976. Econometric measurement of the duration of advertising effects on sales. *Journal of Marketing*, 13: 345–357.

Cramer JS. 1969. *Empirical econometrics*. Amsterdam/London: North-Holland Publishing Co.

Cramer JS. 1978. On prediction. *Proceedings Bicentennial Congress Wiskundig Genootschap*, Amsterdam: MC tract serie, 100–101, 123–132.

Fase MMG. 1969. *An econometric model of age-income profiles*. Rotterdam: Rotterdam University Press.
Fase MMG. 1971. On the estimation of lifetime income. *Journal of the American Statistical Association*, 66: 686–692.
Fase MMG. 1972. The distribution of lifetime earnings: a problem of intertemporal aggregation. *Statistica Neerlandica*, 26: 103–111.
Fase MMG. 1981. *Op het breukvlak van micro- en macro-economie*. Leiden: Stenfert Kroese.
Granger CWJ. 1969. Investigating causal relations by econometric models and cross-spectral methods. *Econometrica*, 37, 3: 424–438.
Guilkey DK and Salemi MK. 1982. Small sample properties of three tests for Granger-causal ordering in a bivariate stochastic system. *Review of Economics and Statistics*, 67, 4: 668–680.
Hanssens DM. 1980. Bivariate time-series analysis of the relationship between advertising and sales. *Applied Economics*, 12: 329–339.
Helmer RM and Johansson JK. 1977. An exposition of the Box-Jenkins transfer function analysis with an application to the advertising-sales relationship. *Journal of Marketing Research*, 14, 227–239.
Hsiao CH. 1979. Autoregressive modelling of Canadian money and income data. *Journal of the American Statistical Association*, 74: 553–560.
Hsiao CH. 1982. Autoregressive modelling and causal ordering of economic variables. *Journal of Economic Dynamics and Control*, 4: 243–259.
Jenkins GM. 1974. The interaction between the muskrat and mink cycles in North Canada, in: *Proceedings of the 8th international biometric conference, August 1974 Constanta Rumania*, edited by L.C.A. Corsten and T. Postelnicu. Editura Academici Republicii Socialiste Romania, 55–71.
Kang H. 1981. Necessary and sufficient conditions for causality testing in multivariate ARMA models. *Journal of Time Series Analysis*, 2, 2: 95–101.
Kapoor SG, Madhok P and Wu SM. 1981. Modelling and forecasting sales data by time series analysis. *Journal of Marketing Research*, 18: 94–100.
Maravall A. 1981. A note on identification of multivariate time-series models. *Journal of Econometrics*, 16, 2: 237–247.
Nerlove M, Grether DM and Carvalho JL. 1979. *Analysis of economic time series: a synthesis*. New York: Academic Press.
Osborn DR. 1984. Causality testing and its implications for dynamic econometric models. *Economic Journal*, 94 (supplement), 82–96.
Palda KS. 1964. *The measurement of advertising effects*. Englewood Cliffs: Prentice Hall.
Sims CA. 1972. Money, income and causality. *American Economic Review*, 62:540–552.
Theil H. 1958. *Economic forecasts and policy*. Amsterdam: North-Holland Publishing Co.
Tiao GC and Box GEP. 1981. Modelling multiple time series with applications. *Journal of the American Statistical Association*, 76: 802–816.
Tiao GC and Tsay RS. 1983. Multiple time series modelling and extended sample cross-correlations. *Journal of Business and Economic Statistics*, 1, 1: 43–56.
Umashankar S and Ledolter J. 1983. Forecasting with diagonal multiple time series models: an extension of univariate models. *Journal of Marketing Research*, 20: 58–63.
Wei WWS. 1978. Some consequences of temporal aggregation in seasonal time series models, in: *Seasonal Analysis of Economic Time Series*, edited by A. Zellner. Washington: 433–444.
Zellner A. 1979. Causality and econometrics, in: *The aspects of policy and policymaking*, edited by K. Brunner and A.H. Meltzer. Amsterdam: North-Holland Publishing Co., 9–54.

On the rationale for and scope of regression models in econometrics

JAN F. KIVIET and GEERT RIDDER

1. Introduction

In a recent book Zellner (1982, p. 26) reviews some basic issues in econometrics, past and present. He recalls the discussion between Keynes and Tinbergen in the 1930s on the appropriateness of applying statistical methods to non-experimental economic time-series data, on the possible incorrect specification of relationships, and on the suitability of econometric methods and models for succesfull prediction and policy making; Zellner also briefly mentions a few other controversial basic issues in econometrics, viz. the causal interpretation of econometric models, the appropriateness of probability statements made in econometric analysis, and the claimed success of rather simple time-series models in comparison with elaborate econometric models. On many of these fundamental issues various conflicting opinions are still propounded in the literature. At the same time in most leading textbooks – that according to their titles deal with the foundations, principles, theory and analysis of models, methods and techniques of econometrics – many of the above-mentioned fundamental matters are not treated explicitly at all.

In discussing in this paper some aspects of the rationale for and the scope of the simple linear regression model we touch on various of the basic issues enumerated by Zellner. In the next Section we first address some very general aspects of econometric modelling, viz. the approximations involved in modelling, the stochastic nature of econometric models, and the now quite widely accepted view that a (unique) correct true model never exists. Then in Section 3 we recapitulate the basic logical principles behind the simple regression model as expressed in the literature; we note that there appear to be two diverging approaches towards regression. In Section 4 we argue that one of the two approaches cannot be used very fruitfully in econometrics, especially in the context of non-experimental time-series data. We also list a number of problems with respect to the applicability of the other approach. Then in Section 5 we discuss the correspondence between particular regression-model

R.D.H. Heijmans and H. Neudecker (eds.), The Practice of Econometrics. ISBN 90-247-3502-5.
© 1987, Martinus Nijhoff Publishers, Dordrecht. Printed in the Netherlands.

assumptions on the one hand and characteristics of real-world relationships on the other. Extending an approach originating from Cramer (1973a) we argue that given a true relationship of a particular nature, there is a case for the (linear) regression model. In Section 6 we assume that a practitioner is convinced by these arguments and uses a linear regression model to analyse the relationship between variables. We show that a careless analyst may find a relationship between variables which are completely unrelated. This phenomenon has been discussed earlier by Granger and Newbold (1974, 1977) who coined the term spurious regression. Here we point at some weaknesses in their formulation and suggest a modification. The modified formulation supports the intuition that there is danger in estimating simple regression models using time series that 'grow together'. We also present a spurious regression problem that may upset cross-section regression analysis. This highlights the danger of a pure measurement approach to econometrics, in which we just fit regression models to variables that for economic reasons might be related, and in which little or no attention is paid to the indispensable statistical underpinning of regression results. In the concluding section we stress the necessity to append regression results with validation checks of the model such as misspecification tests.

2. Data generation and econometric modelling

Econometricians generally agree that an econometric model is an analytical and stochastic construct that attempts to mimic the process that actually generated the data under study. An operational model that is to be estimated and analyzed on the basis of a finite number of sample observations usually only provides an approximation to such a data-generating process (DGP); in practice a fully deterministic and completely adequate description of a true economic relationship will be unobtainable, even if one can control all the relevant economic factors. Invariably errors will be involved such as the noise due to measurement and rounding errors, disturbances originating from non-fulfilment of the ceteris-paribus condition, the effect of non-economic factors, and perhaps deflections resulting from aggregation of the relevant data over individual agents or over time. All these disturbing factors, which are usually represented by stochastic error terms, are an inherent part of econometric models.

For much the same reasons stochastic disturbances could perhaps also be seen as intrinsic elements of DGPs. It is our opinion that it is irrelevant for the foundation of econometric modelling whether or not a unique and exact deterministic representation of the DGP is supposed to exist for each and every phenomenon, and likewise whether or not such a representation is (at least in theory) thought to be obtainable. A DGP is better considered merely as a hypothetical concept. We do believe, however, that for the rationalization of stochastic econometric models, and for judgements on their adequacy it is inevitable to adopt an axiom like the following: 'For the (set of) economic relationship(s) under study a whole range of data-generating representations (DGRs) is conceivable. These DGRs are (in principle obtainable) econometric models and each of these could possibly have generated the data; i.e. the actual data and data generated according to the DGRs are observationally equivalent'. For particular rather simple relationships some of these DGRs may come close to a (genuine one and only fully detailed and deterministic) DGP. However, most DGRs will be much simpler and more an abstraction of reality although (if available) each of them may be used as an operational and adequate model. This quality of DGRs is achieved by representing the approximation errors in the deterministic part by stochastic factors. The view expressed here implies that for one relationship, different models – usually varying in detail – may be adequate, though not all will be equally efficient with respect to specific types of analysis.

In the next Section we review two general approaches that can be found in the literature which lead up to the presentation of the regression model. This precedes a discussion of the justification of the regression model for application in the field of econometrics, where we pay special attention to the presentation by Cramer. We find his line of thought very illuminating, while, on the other hand many other texts on the foundations of the regression model lack convincing arguments for choosing to model economic phenomena in this way. Usually the regression model and its formal analysis under various model assumptions is just presented as such, without much reference to the nature of actual data-generating phenomena and its correspondence with the particular regression model assumptions.

3. Two approaches towards regression

It is not our intention to give here a historical overview of the origins and later developments of the regression model in the statistical and econometric literature. Nor shall we attempt to give credit to the various authors who have contributed to the present well-established position of this type of model. We will only give a review of (and some comments on) the two predominant ways in which the regression model is usually presented. Next we will assess the empirical content of the model: we will investigate the applicability of the various model assumptions to practical situations of interest in econometrics.

The two approaches that are usually followed to introduce the regression model have been indicated by Wold (1953, p. 205) by the terms 'Galton-Yule specification' and 'Gauss-Fisher specification'. These involve the following (proofs of the statements can be found in the literature; see, for instance, Malinvaud (1964, p. 101)):

(A) *Galton-Yule* (regression in multivariate distributions)
Let $Z = [z_1, \ldots, z_N]^T$ be an NxK matrix, where Z contains a sample of N drawings of the stochastic K-element vector z. We assume that

(A.1) The vectors z_n, $n = 1, \ldots, N$ are identically distributed with finite non-singular covariance matrix.

(A.2) The vectors z_n, $n = 1, \ldots, N$ are mutually independent.

Now let y_n indicate an arbitrary element of the vector z_n for $n = 1, \ldots, N$ and let the (K-1)x1 vector x_n contain all the other elements of z_n. We use g(x) to denote the conditional expectation of y given x. The relationship

$$y = g(x), \text{ with } g(x) = E(y|x) \tag{1}$$

is defined as the *'theoretical regression'*, or the *'regression curve'* – see Cramèr (1945, p. 270) – of the regression of y on x. The theoretical regression g(x) is closest to drawings y in the least-squares sense since among all functions f(x) the minimum of $E\{y-f(x)\}^2$ is found for $f \equiv g$. This follows from

$$E\{y - f(x)\}^2 = E_x[E\{(y - f(x))^2|x\}]$$

while the latter is minimized by taking $f(x) = E(y|x) = g(x)$.

In practice the distribution of y given x and hence the function g may be unknown and min $E\{y-f(x)\}^2$ might be taken only for functions f that are linear in x. Then we obtain the *'best linear predictor'* of y or the *'theoretical linear regression'* which is defined as the hyperplane

$$y = \alpha_1 + x^T \alpha_2 \tag{2}$$

where α_2 is a (K-1)x1 vector of constants and α_1 is a constant scalar with

$$\alpha_2 = [\text{cov}(x,x)]^{-1}\text{cov}(x,y); \qquad \alpha_1 = Ey - Ex^T\alpha_2. \tag{3}$$

Here cov(v,w) indicates $E(v-Ev)(w-Ew)^T$ for general column vectors v and w. In (3) the matrix cov(x,x) is non-singular due to (A.1). Substitution of (3) into (2) leads to

$$y = Ey + (x - Ex)^T[\text{cov}(x,x)]^{-1}\text{cov}(x,y). \tag{4}$$

If we assume, next to (A.1) and (A.2) that
(A.3) The theoretical regression of y on x is linear,
then the theoretical regression $y = g(x)$ and the best linear predictor (4) coincide. If we assume, next to (A.1) and (A.2) that

(A.4) z_n, $n = 1, \ldots, N$ has a multivariate normal distribution

then assumption (A.3) is in fact redundant as it is implied by (A.4) and (A.2).

We now consider linear least-squares estimation in this model. Let Y denote the N-element column vector of sample values on y, hence $Y = (y_1, \ldots, y_N)^T$ and further, let $X_2 = [x_1, \ldots, x_N]^T$ and $X = [1 : X_2]$. Here X is an NxK matrix where the first column has all elements equal to unity. The ordinary least-squares (OLS) vector is given by

$$a = (X^T X)^{-1} X^T Y. \tag{5}$$

This contains the K coefficients of the *'empirical linear regression'* of y on x and a constant. Under assumptions (A.1) and (A.2) we have

$$\plim_{N \to \infty} a = \alpha = (\alpha_1, \alpha_2^T)^T. \tag{6}$$

If, additionally, (A.3) is valid then we also have

$$E(a|X) = \alpha \text{ and hence } Ea = E_x E(a|X) = \alpha, \tag{7}$$

so then the empirical regression is unbiased for the theoretical regression. Furthermore, we then have

$$\cov(a|X) = \var(y|X) \cdot (X^T X)^{-1}. \tag{8}$$

If, in addition, (A.4) is valid then it can be found that the conditional distribution of the quadratic form

$$\frac{N-K}{r} \frac{(Ra - R\alpha)^T [R(X^T X)^{-1} R^T]^{-1} (Ra - R\alpha)}{Y^T [I - X(X^T X)^{-1} X^T] Y}, \tag{9}$$

where R is a rxK matrix with rank $(R) = r \leq K$, is simply $F_{r, N-K}$ and hence independent of the conditioning variables. This enables the testing of hypotheses on the value of $R\alpha$, although Ra is non-normal.

We shall show now that the assumptions (A.1) through (A.4) can be stated in another equivalent way. To that end we introduce the stochastic (usually unobservable) variable

$$u = y - E(y|x) = y - \alpha_1 - x^T \alpha_2. \tag{10}$$

From (A.1) through (A.4) it follows that drawings of this variable are mutually independent, and are identically normally distributed with finite variance; we find

$$Eu = E_x E(u|x) = E_x 0 = 0$$

and upon using (3) we find

$$\begin{aligned}\var(u) &= \var(y) - 2\cov(y,x)\alpha_2 + \alpha_2^T \cov(x,x)\alpha_2 \\ &= \var(y) - \cov(y,x)[\cov(x,x)]^{-1}\cov(x,y) \\ &= \sigma_u^2, \text{ a constant.}\end{aligned}$$

Moreover, we obtain
$$\begin{aligned}\operatorname{cov}(x,u) &= E(x - Ex)(y - \alpha_1 - x^T\alpha_2)\\ &= \operatorname{cov}(x,y) - \operatorname{cov}(x,x)\alpha_2\\ &= \operatorname{cov}(x,y) - \operatorname{cov}(x,x)[\operatorname{cov}(x,x)]^{-1}\operatorname{cov}(x,y)\\ &= 0,\end{aligned}$$

hence u is seen to be independent of x. These results can be summarized as follows:

(A*.1) $y = \alpha_1 + x^T\alpha_2 + u$
where α_1 and α_2 are fixed and u is an (unobservable) random variable with $Eu = 0$ (otherwise α_1 is unidentified).
(A*.2) $\operatorname{var}(u) = Eu^2 = \sigma_u^2$ is finite and constant.
(A*.3) drawings of u are mutually independent.
(A*.4) x and u are distributed independently.
(A*.5) x is a (K-1)-variate normally distributed vector with finite mean and non-singular covariance matrix; drawings of x are mutually independent.
(A*.6) u has a normal distribution.

As (A*.1) through (A*.6) are implied by (A.1) through (A.4), while (A*.1) through (A*.6) imply (A.1) through (A.4), these two sets of assumptions are equivalent. Before we discuss the relevance of the Galton-Yule approach we first formulate the regression model according to the second approach which Wold calls:

(B) *Gauss-Fisher* (reduced form causal relationships).

This specification (which appears in some form or another in all textbooks on econometrics) is based on a (causal) relationship which specifies that a dependent variable, say y, is determined by a number of explanatory variables, say a vector x, and by other (unobservable) elements according to a linear formula that holds for all observations. We shall discuss a few closely related versions of this specification. The initial set of assumptions is as follows:

(B.1) $y_n = \beta_1 + x_n^T\beta_2 + \varepsilon_n \quad n = 1, \ldots, N$
where β_1 and β_2 are fixed, both x_n and β_2 are (K-1) element vectors, and ε_n is a random variable (the disturbance term) with $E\varepsilon_n = 0$.
(B.2) $\operatorname{var}(\varepsilon_n) = E\varepsilon_n^2 = \sigma_\varepsilon^2$ is finite and constant.
(B.3) ε_n and ε_m are mutually uncorrelated for $n \neq m$.
(B.4) the distribution of ε_n does not depend on x_m, $m = 1, \ldots, N$ for all $n = 1, \ldots, N$ (x is *exogenous* with respect to ε).
(B.5) x varies in such a way that (i) if $N > K$ there is no extreme multicollinearity, and (ii) for $N \to \infty$ consistent estimators of β_1 and β_2 can be obtained.

The notation we have used enables the whole sample of N observations to be denoted again by the Nx1 vector Y and the NxK matrix X. However, to stress the distinction between the specifications (A) and (B) we denote the unobservables here by $\beta = (\beta_1, \beta_2^T)^T$ and $\varepsilon = (\varepsilon_1, \ldots, \varepsilon_N)^T$ instead of α and $(u_1, \ldots, u_N)^T$. Hence, we may write

$$Y = X\beta + \varepsilon \qquad (11)$$

and for the OLS coefficient estimator we now have

$$b = (X^TX)^{-1}X^TY. \qquad (12)$$

Under (B.1) through (B.5) we find

$$Eb = \beta; \quad cov(b|X) = \sigma_\varepsilon^2 (X^TX)^{-1}$$

and

$$\plim_{N \to \infty} b = \beta \qquad (13)$$

If we have additionally

(B.6) ε_n is normally distributed

then we find

$$b|X \sim N(\beta, \sigma_\varepsilon^2 (X^TX)^{-1}). \qquad (14)$$

As the conditional distribution of Rb, with R as in (9), is normal it is easy to prove that the statistic

$$\frac{N-K}{r} \frac{(Rb - R\beta)^T[R(X^TX)^{-1}R^T]^{-1}(Rb - R\beta)}{Y^T[I - X(X^TX)^{-1}X^T]Y} \qquad (15)$$

has a $F_{r, N-K}$ distribution, which is independent of X. Statistic (15) can be used to test hypotheses on the value of $R\beta$.

If, instead of (B.4), we have

(B*.4) x_n and ε_m are distributed independently for $n \leq m$ (x is *predetermined* with respect to ε) for $n,m = 1, \ldots, N$

then we find

$$b \sim AN(\beta, \sigma_\varepsilon^2 (X'X)^{-1}), \qquad (16)$$

where AN indicates: approximately normal according to considerations involving asymptotics.

Upon comparison of the (A) and (B) specification we see that there is some correspondence between the set of assumptions (B.1) through (B.5) and the set consisting of (A.1) through (A.3). Also when both (B.6) and (A.4) are added the two approaches have some elements in common.

The main differences are:
(i) in (A) the x vectors are mutually independent, whereas they are largely arbitrary in (B),
(ii) in (A) the choice of y and x variables (from z) is arbitrary, whereas in (B) they have to represent a 'true' causal relationship, which has to be specified quite precisely, because
(iii) the remaining factors (the disturbances ε) have to obey rather strict characteristics, whereas the corresponding random factors u of (A) will have comparable characteristics more or less automatically due to the far-reaching assumptions on the distribution of the vector z.

4. The relevance of the two regression approaches for econometrics

We now come to a discussion of the applicability of the regression model and hence of the relevance of the two sets of assumptions for economic data. For the Galton-Yule specification (A) in particular we have to conclude that the assumptions will seldom be justifiable; at the same time regression results based on this approach may be rather impracticable as we shall see.

For economic time-series data assumption (A.1) on the mutual independence of observations at different points in time will almost never be satisfied. On the other hand cross-section data that satisfy both the assumptions (A.1) and (A.2) may be obtained from a random sample of a population. Then, however, it will usually be questionable whether or not the elements in the vector z have a normal distribution or – more basically – whether the theoretical regression is linear. If it is not, then either the variables have to be transformed to achieve linearity or normality, or one has to consider non-linear least-squares regression. In the latter case we have to face the problem of parameterizing $g(x) = E(y|x)$. If the data transformation or the regression parameterization is performed inappropriately, then the empirical (non-)linear regression will in general be biased for the theoretical regression.

Apart from the difficulties in meeting the requirements of the assumptions, a disquieting element in approach (A) is that, if the assumptions are valid, regression may also be performed using only some subset of the conditioning variables x. At the same time the addition of mutually independent normally distributed conditioning variables would also be allowed. However, in both cases the regression curve changes when the sets of regressors involved are correlated (which is usually the case). Therefore, the interpretation of the regression coefficients is rather difficult in approach (A); hence, it is not obvious what purpose a Galton-Yule exercise in coefficient estimation and curve fitting is supposed to serve. Yet another difficulty in a Galton-Yule based regression is that conditional forecasts, or conditional extra – or intrapolation cannot be performed in a way economists are familiar with. The problem is that in forecasting exercises in the (A)-framework it seems that one is not free to choose the values of the conditioning variables x; after all these regressor values have to obey the characteristics of drawings from the distribution of x.

On the whole we have to conclude that the Galton-Yule approach is of little significance to econometrics – see also Malinvaud (1964, p. 104) – and the standard is the Gauss-Fisher approach as presented in all econometrics textbooks. In the course of the development of econometric theory many modifications and generalizations of the set of assumptions (B) as stated in the foregoing section have been suggested. For instance, regression techniques have been developed to deal with relationships that are non-linear in the regression coefficients. Then, instead of assumption (B.1), we might have

(B*.1) $y_n = g(x_n, \beta) + \varepsilon_n$ $n = 1, \ldots, N$
with β a fixed K-element vector of unknown parameters for the function g which is completely specified; ε_n is a random disturbance term with $E\varepsilon_n = 0$.

Also extensions with respect to the covariance matrix of ε and concerning the joint distribution of x and ε have been put forward. Apart from that, techniques have been developed for systems of (simultaneous) relationships. All these modifications still obey basic characteristics of the (B) approach: viz. they all start off from the specification of a (set of) 'cause-effect' relationship(s) between a (set of) regressand(s) and a number of regressors plus a stochastic disturbance term, specifying also a number of characteristics of the distribution of the disturbance(s).

Although much effort has been put in developing techniques to analyse data that are supposed to obey (B)-type assumptions, and many tests have been designed to check whether data conform to a particular set of such assumptions or some specific generalization of this set of (B)-type assumptions, the highly relevant question of whether (B)-type asumptions as such are, or may be, justifiable for economic data is seldomly posed. Usually it is simply stated that the data are actually generated according to the (B)-framework; then in fact the real world is subordinated to the model, while it should be the other way around. For instance, in the recent Handbook of Econometrics (1983, 1984), notably in the chapters 3, 6 and 7, various (regression) models are presented and analyzed by stating sets of assumptions, while the relevance of these models for econometrics is presupposed.

Whether the (B)-type regression model can be adequate for use in practice is examined in a rather non-standard way in Cramer (1973a). He tries to demonstrate how model assumptions relate to reality, and he investigates the type of approximations that are involved in modelling a relationship by a linear regression model. Here we will only paraphrase his line of thought (the reader is referred to the original and lucid pages 76 through 83 of Cramer's 'Empirical Econometrics'), and in the next section we will give a generalization of this approach.

Starting off from the premise that all phenomena are completely determined by causal relationships Cramer says that any economic variable y is determined by a (great) number, say M, of variables, giving

$$y = \varphi(x_1, \ldots, x_M).$$

For N (hypothetical) observations on the variables involved we then have

$$y_n = \varphi(x_{n1}, \ldots, x_{nM}), \qquad n = 1, \ldots, N. \tag{17}$$

Expanding φ in a Taylor series around the point

$$(\bar{x}_1, \ldots, \bar{x}_M), \text{ where } \bar{x}_i = \frac{1}{N} \sum_{n=1}^{N} x_{ni}, \qquad i = 1, \ldots, M$$

and assuming appropriate differentiability of φ, we obtain

$$y_n = \varphi(\bar{x}_1, \ldots, \bar{x}_M) + \frac{\partial f}{\partial x_1}(x_{n1} - \bar{x}_1) + \ldots + \frac{\partial f}{\partial x_M}(x_{nM} - \bar{x}_M)$$
$$+ \frac{1}{2!}\left\{\frac{\partial^2 \varphi}{\partial x_1^2}(x_{n1} - \bar{x}_1)^2 + 2\frac{\partial^2 \varphi}{\partial x_1 \partial x_2}(x_{n1} - \bar{x}_1)(x_{n2} - \bar{x}_2) + \ldots\right\} \tag{18}$$
$$+ \ldots, \qquad n = 1, \ldots, N$$

where all derivatives are evaluated at $(\bar{x}_1, \ldots, \bar{x}_M)$. It is quite likely that many of the M determining factors of y are constant over the N observations. Let this be the case for M-L of them so that we have

$$x_{nj} - \bar{x}_j = 0, \text{ for } j = L+1, \ldots, M; n = 1, \ldots, N. \tag{19}$$

Upon substitution of (19) in (18) all terms involving (partial (higher order)) derivatives with respect to x_j, $j = L+1, \ldots, M$ vanish.

We now define the constants

$$\beta_1 = \varphi(\bar{x}_1, \ldots, \bar{x}_M) - \sum_{k=1}^{K-1} \bar{x}_k \frac{\partial \varphi}{\partial x_k}\bigg|_{(\bar{x}_1, \ldots, \bar{x}_M)} \tag{20}$$

and

$$\beta_{k+1} = \frac{\partial \varphi}{\partial x_k}\bigg|_{(\bar{x}_1, \ldots, \bar{x}_M)} \qquad k = 1, \ldots, K-1.$$

Further, we introduce

$$x_{no} \equiv 1, \quad n = 1, \ldots, N$$

and we can then rewrite (18) as

$$y_n = \sum_{k=1}^{K} \beta_k x_{n, k-1} + r_n, \quad n = 1, \ldots, N \tag{21}$$

whereby we have implicitly defined the remainder term r_n. It equals the sum of all the terms of (18) involving (partial) derivatives with respect to x_j, $j = K, \ldots, L$ and all terms involving higher-order derivatives with respect to x_k, $k = 1, \ldots, L$. In (21) only the first $K-1$ variables of $(x_1, \ldots, x_M)$ are explicitly taken into account in linear form; all other non-zero terms of (18), except β_1, are collected in the remainder term. Notice that (21) still represents the exact deterministic relationship.

We are now in a position where we can discuss the relevance of the regression-model assumptions according to the approach (B) for application to real-world relationships like (21). The basic issue is reduced now to the question whether the remainder term r_n can be modelled adequately by a random disturbance term obeying the required characteristics.

In answering this question we face various problems, such as:
(i) the above analysis is only applicable for functions φ that are differentiable;
(ii) the above analysis is only applicable for variables x_j, $j = 1, \ldots, M$ that vary continuously whereas many economic relationships involve qualitative variables (sex, type of education, etc.) or other discrete variables (number of children, number of cars, etc.);
(iii) in the above analysis the variables $x_1, \ldots, x_{K-1}$ all vary independently (they are 'variation-free'), hence polynomial regressions, interaction terms, and other linearized non-linear relationships are excluded here, although Cramer is well-aware of their importance, see Cramer (1973b);
(iv) it is not obvious how one can justify non-linear regressions by the linear result (21);
(v) the approach has been criticized by White (1980) on the grounds that the linear least-squares regression coefficients of (21) will generally diverge from the coefficients (20). This is immediately understood by observing that the least-squares regression plane intersects with the point $(\bar{y}, \bar{x}_1, \ldots, \bar{x}_{K-1})$, and not necessarily with $(\varphi(\bar{x}_1, \ldots, \bar{x}_{K-1}), \bar{x}_1, \ldots, \bar{x}_{K-1})$, as $\bar{y}$ and $\varphi(\bar{x}_1, \ldots, \bar{x}_{K-1})$ usually don't coincide for non-linear functions φ.
(vi) the remainder r_n of (21) is deterministic so it might be inappropriate to model it by a random variable;
(vii) in practice we will not have observations available on all variables $x_K, \ldots, x_M$ so we cannot check the characteristics of r_n.

In order to deal with these problems, we will first generalize Cramer's analysis in some respects.

5. A justification for (linear) regression in econometrics

We generalize Cramer's assumption that all economic phenomena are completely determined by deterministic causal relationships, and assume that y is determined by an unspecified function φ of M variables $v_1, \ldots, v_M$ and a stochastic disturbance term v, according to

$$y = \varphi(v_1, \ldots, v_M) + v. \tag{22}$$

This relationship in fact depicts the DGP of y which is deterministic (non-stochastic) if v's distribution is degenerate. We suppose that given the information in $\varphi(v_1, \ldots, v_M)$ drawings from v will be mutually stochastically independent, have expectation zero and have finite variance. The variables y and $v_1, \ldots, v_M$ may represent (transformations of) economic and non-economic variables. However, we assume that $v_1, \ldots, v_M$ are 'variation-free': they can vary independently, although they perhaps do not in a particular sample. Expression (22) can be justified by noting that it is conceptually possible to divide the determining factors of y into two groups which refer to physically unrelated mechanisms e.g. the economic mechanism described by φ and the measurement mechanism described by v. If we are interested in φ we can treat v as a random variable with the indicated properties and the distribution of v does not depend on $v_1, \ldots, v_M$. Treating v as a random variable is just a matter of convenience: it simplifies the description of the DGP considerably.

The variables $v_1, \ldots, v_M$ may be either discrete (finite number of possible values) or continuous. We fix the discrete variables at particular values and consider the function φ as a function of the continuous variables (different for all choices of the discrete variables). For a particular choice of the discrete variables the domain of φ may contain points where φ is not differentiable to arbitrary order. We assume that these points are located in such a way that we can partition the domain of φ into a number of disjoint subsets where the points of non-differentiability are on the boundary of the disjoint subsets. Hence, we divide the domain of $v_1, \ldots, v_M$ into sub-domains with the property that on a sub-domain φ is differentiable to arbitrary order with respect to the continuous variables. Note that on a sub-domain the discrete variables are constant.

Let us now concentrate on one arbitrary sub-domain and suppose that we have N observations, then we may write

$$y_n = \varphi(v_{n1}, \ldots, v_{nM}) + v_n, \quad n = 1, \ldots, N. \tag{23}$$

Let M-L of the variables $v_{n1}, \ldots, v_{nM}$ be constant over the sample in this particular sub-domain, either due to the actual construction of this domain (the discrete variables) or directly due to the values obtained in sampling. We therefore have

$$v_{nj} = \tilde{v}_j \quad \text{for } j = L+1, \ldots, M; n = 1, \ldots, N$$

which reduces (23) to

$$y_n = \varphi(v_{n1}, \ldots, v_{nL}, \tilde{v}_{L+1}, \ldots, \tilde{v}_M) + v_n. \tag{24}$$

From the arguments given above it follows that Taylor's theorem can be applied to (24). Upon expansion in a point $(\tilde{v}_1, \ldots, \tilde{v}_L, \tilde{v}_{L+1}, \ldots, \tilde{v}_M)$ of the sub-domain we obtain

$$y_n = \varphi(\tilde{v}_1, \ldots, \tilde{v}_M) + \frac{\partial f}{\partial v_1}(v_{n1} - \tilde{v}_1) + \ldots + \frac{\partial f}{\partial v_L}(v_{nL} - \tilde{v}_L)$$

$$+ \frac{1}{2!}\left\{\frac{\partial^2 \varphi}{\partial v_1^2}(v_{n1} - \tilde{v}_1)^2 + 2\frac{\partial^2 \varphi}{\partial v_1 \partial v_2}(v_{n1} - \tilde{v}_1)(v_{n2} - \tilde{v}_2) + \ldots\right\}$$

$$+ \ldots + v_n. \tag{25}$$

Apart from the first and the last term on the right-hand-side of (25), the various terms involve (partial) derivatives of φ which can be divided into three categories. Firstly there is a group of terms which are zero because the derivatives involved are zero; then the respective variables v_j or products of various v_j's do not influence y_n, and they may be omitted. Secondly, there is a number of terms, say K-1, we are willing to take explicitly into account. Let we indicate these by

$$\beta_2 x_2 + \ldots + \beta_K x_K \tag{26}$$

where the β_j denote partial (higher order) derivatives of φ in $(\tilde{v}_1, \ldots, \tilde{v}_M)$, and the x_j denote the corresponding (powers or products of) variables v. Notice that K-1 may exceed L and that we can have, for instance, $x_3 = v_2^2$ or $x_4 = v_2 \cdot v_3$. Hence, the variables $x_2, \ldots, x_K$ are not necessarily 'variation-free' and the linear combination of x's in (26) may be non-linear in the v's. Finally, there are the non-zero terms that are collected in a remainder term, and hence we have

$$y_n = \varphi(\tilde{v}_1, \ldots, \tilde{v}_M) + \beta_2 x_{n2} + \ldots + \beta_K x_{nK} + r_n + v_n$$

$$= \sum_{k=1}^{K} \beta_k x_{nk} + \tilde{r}_n, \qquad n = 1, \ldots, N \tag{27}$$

where

$$\beta_1 = \varphi(\tilde{v}_1, \ldots, \tilde{v}_M) + \frac{1}{N}\sum_{n=1}^{N} r_n,$$

$x_{n1} \equiv 1$, and

$$\tilde{r}_n = (r_n - \frac{1}{N}\sum_{n=1}^{N} r_n) + v_n.$$

To indicate that this equation only concerns a sub-domain, we slightly adapt the notation now. Let we have D domains, and $N^{(d)}$ observations and $K^{(d)}$

regressors in the equation for domain $d = 1, \ldots, D$. Then we modify (27) in

$$y_n^{(d)} = \sum_{k=1}^{K(d)} \beta_k^{(d)} x_{nk}^{(d)} + \tilde{r}_n^{(d)} \qquad n = 1, \ldots, N^{(d)}; d = 1, \ldots, D. \qquad (28)$$

We can now rewrite (28) as a regression equation with $K^* = \sum_d K^{(d)}$ regressors for $N^* = \sum_d N^{(d)}$ observations, giving

$$y_i^* = \sum_{j=1}^{K^*} \beta_j^* x_{ij}^* + \tilde{r}_i^*, \qquad i = 1, \ldots, N^* \qquad (29)$$

where for $d = 1, \ldots, D$ and $n = 1, \ldots, N^{(d)}$ we have $i = n + \sum_{h=1}^{d-1} N^{(h)}$ and $j = k + \sum_{h=1}^{d-1} K^{(h)}$ with

$$y_i^* = y_n^{(d)}; \ \tilde{r}_i^* = \tilde{r}_n^{(d)}; \ \beta_j^* = \beta_k^{(d)}$$

and

$$x_{ij}^* = \delta_i^{(d)} x_{nk}^{(d)}$$

where

$$\delta_i^{(d)} = \begin{cases} 1 & \text{if } \sum_{h=1}^{d-1} N^{(h)} < i \leq \sum_{h=1}^{d} N^{(h)}. \\ 0 & \text{otherwise.} \end{cases}$$

We see that even for non-differentiable functions φ and discrete causal variables v the linear regression equation with a remainder term can be obtained. If some of the coefficients have the same values for different sub-domains, then (29) can be simplified; the number of regressors K^* can be reduced, but the general characteristics of the linear regression equation are maintained, and the remainder term $\tilde{r}_i^*$ is unaltered.

The results obtained above solve the problems (i) through (iii) listed at the end of the foregoing section. With respect to problem (iv) we can remark that now our DGP (22) is in fact a non-linear regression, which of course may (if possible) be estimated directly. Alternatively it may be simplified in another non-linear regression, or perhaps be approximated by a linear regression of the form (29). Anyhow, if linear regressions can be justified (see below), then so can (well-specified) non-linear regressions. With respect to (v) and the criticism of White we can observe that in the above generalized analysis the point(s) of expansion remain(s) unspecified. Hence, we can always choose the point of expansion such that at least the partial derivatives of φ with respect to the regressors coincide with the least-squares probability limits. In addition, we have to point out that Cramer's and the above generalized analysis serve a purpose completely different from White's (1980, p. 152). The latter considers regression in the Galton-Yule framework and presupposes validity of the model assumptions (A.1) and (A.2). He does not assume normality nor linearity, and hence – compare (10) – he has

$$y = E(y|x) + u = g(x) + u$$

and examines the properties of linear least squares aproximations of $g(x)$ and mentions result (6). We, on the other hand, consider a DGP (where White considers a model) of the (non-linear regression) form

$$y = \varphi(v_1, \ldots, v_M) + v,$$

where neither y nor $v_1, \ldots, v_M$ are i.i.d. random variables and where perhaps many of the variables v are unobservable. For that case we attempt to demonstrate that the Gauss-Fisher type model assumptions may be applicable. For this matter we see from (29) that for such DGPs a linear regression equation can be obtained that conforms to the one in assumption (B.1) but with the following difference: in (29) we have a remainder term $\bar{r}^*$ whereas in (B.1) there is a stochastic disturbance term ε with zero mean, obeying the assumptions (B.2) through (B.4) and eventually (B.6). From (27) it follows that the remainder term has a deterministic component and – like in problem (vi) – we have to question whether this can be modelled adequately by a random variable. On this issue we like to quote Cramer (1971, p. 79): 'As a matter of fact there is no intrinsic difference between stochastic and deterministic phenomena; even dice behave according to the laws of mechanics. A variable is random merely because we chose, as a matter of convenience, to describe its behaviour by a probability distribution'. In line with these thoughts we argue that it is by the very act of replacing the remainder $\bar{r}^*$ of (29) by a stochastic variable that we on the one hand only produce a model (an approximative representation) of the true relationship (which usually is inaccessible itself for analysis in its full detail), but that on the other hand we keep the option to obtain a sufficiently accurate representation of the relationship since we do not neglect the omitted terms altogether.

If for a DGP a (linear) regression approximation can be found such that the DGP can be modelled effectively by the Gauss-Fisher type DGR (data-generating representation, see section 2), then this specification of the regression equation (the actual choice of the regressors $x_2, \ldots, x_K$) is not necessarily unique. If the Gauss-Fisher approach is at all applicable it is quite possible that a whole range of adequate DGRs exists: the crucial requirement is that the remainder term can be represented by an observationally equivalent stochastic disturbance term, which obeys particular (B)-type assumptions. Because of problem (vii) it is impossible to check beforehand whether the remainder $\bar{r}^*$ has the appearance of a stochastic variable with the characteristics of ε. For this reason there has been an enormous emphasis in recent development of econometric methods and methodology on (mis)specification testing: after the data have been analyzed on the basis of particular provisionally maintained model assumptions, we check to see whether the results obtained are in fact in agreement with these provisional assumptions. More about this, and many references to literature on testing the validity of the assumptions can be found, for instance, in Pagan (1984) and Kiviet and Phillips (1986) and notably for assumption (B.4) in Ruud (1984). Obviously the independence of regressors and disturbances requires the orthogonality of r_n in (27) with respect to the chosen regressor variables in (26). Through randomization such orthogonality can be obtained by construction; otherwise it can only be achieved through a vigilant misspecification analysis of the regression equation, followed – if necessary – by an effective reformulation of the regression specification. The actual correspondence of the characteristics of the unobservable remainder term with a stochastic disturbance term which does obey the (B)-type assumptions determines the reliability of the inference from regression analysis. This is illustrated in the next section.

6. The spurious regression problem

In the previous section we argued that an (empirical) relation between economic variables can be described by a linear regression model, and we derived some results on least-squares estimation of the model under the Gauss-Fisher type assumptions (B.1) through (B.5) in section 3. In this section we discuss the importance of seemingly technical assumptions as (B.2), (B.3) and (B.4). In particular, we show that neglecting these assumptions i.e. allowing linear regression models that violate these assumptions may lead to erroneous inference on the relation between variables. Transforming or respecifying the model so that it conforms to (B)-type assumptions protects against these errors, as is illustrated by some simple examples in this section.

Consider two (random) variables x and y and assume that the observations x_n, y_n for $n = 1, \ldots, N$ have been generated by

$$x_n = \lambda + x_{n-1} + \eta_n,$$
$$y_n = \mu + y_{n-1} + \zeta_n. \tag{30}$$

The starting values x_0, y_0 are non-stochastic constants. The variables x and y follow a random walk with drift. A random walk with drift models a linearly growing time-series, as can be seen from

$$x_n = \lambda n + \sum_{s=1}^{n} \eta_s + x_0. \tag{31}$$

Note that the variance of x_n also grows linearly. If x_n refers to the logarithm of some variable we have a model for an exponentially growing time-series. Note also that η_n, ζ_n can be interpreted as random shocks to the growth rate in period n. We assume that these shocks are correlated:

$$\zeta_n = \gamma \eta_n + \varepsilon_n. \tag{32}$$

The ε_n in (32) are i.i.d. with constant variance, and are independent of $\eta_1, \ldots, \eta_N$. In the following y is the dependent variable and x is the regressor. Time-series such as x_n and y_n, $n = 1, \ldots, N$ could be observed in economies that experience stable growth.

From (32) it follows that x and y are related. We have

$$\triangle y_n = (\mu - \gamma \lambda) + \gamma \triangle x_n + \varepsilon_n \tag{33}$$

and (33) satisfies the assumptions (B.1) through (B.5). Now let us assume that we started off with the following simple linear regression model

$$y_n = \beta_1 + \beta_2 x_n + \xi_n. \tag{34}$$

Then it can be shown (see the Appendix) that if $\lambda \neq 0$

$$\plim_{N \to \infty} b_2 = \frac{\mu}{\lambda}, \tag{35}$$

where b_2 is the OLS estimator of β_2. Hence, if N is large the OLS estimator equals the ratio of the growth rates. Note that the true nature of the relation between y and x is totally obscured: the probability limit does not depend on γ. Hence (35) holds even if $\gamma = 0$, i.e. if x_n and y_n are stochastically independent. If N is large we can approximate the OLS estimator of β_1 by

$$b_1 \approx \frac{1}{N} \sum_{n=1}^{N} \sum_{s=1}^{n} \zeta_s - \frac{\mu}{\lambda} \frac{1}{N} \sum_{n=1}^{N} \sum_{s=1}^{n} \eta_s. \tag{36}$$

It is shown in the Appendix that although $E(b_1) = 0$, b_1 does not converge to any value if $N \to \infty$. Substituting (35) and (36) in the model we see that if N is large the OLS residuals can be approximated by

$$\hat{\xi}_n \approx \sum_{s=1}^{n} (\zeta_s - \frac{\mu}{\lambda} \eta_s) - \frac{1}{N} \sum_{n=1}^{N} \sum_{s=1}^{n} (\zeta_s - \frac{\mu}{\lambda} \eta_s). \tag{37}$$

Note that the sample mean of the $\hat{\xi}_n$ is 0. It is obvious that the disturbances in (34) do not satisfy assumptions (B.2) through (B.4). Hence, the analyst can discover the erroneous inference by studying the residuals.

The appropriate action in this case is taking first differences while retaining the constant term. In model (33) we obviously find a consistent estimator for the slope coefficient and the true nature of the relation is revealed. From (35) it follows that there is another way to discover the misspecification. If the growth rates μ and/or λ change, the probability limit of b_2 changes. Hence, a non-constant slope coefficient (over different sub-periods) may point to the presence of the indicated problem.

The problem discussed above is, of course, closely related to the problem of spurious regression considered by Granger and Newbold (1974, 1977). The only difference between their assumptions and ours is that they consider the special case where $\mu = \gamma = 0$ in (30) and $\lambda = 0$ in (32). It can be shown (see the Appendix), that under these assumptions we have

$$E(b_2) = 0 \tag{38}$$

which, at first sight, will not give rise to spurious results in (34). On the other hand b_2 does not have a probability limit. Therefore, b_2 does not settle at a particular value in largue samples and this would reveal the spurious nature of the regression. Because the coefficient of determination R^2 does not converge to 0 in this case either, and hence $E(R^2)>0$ for all sample sizes, we find that Jensen's inequality implies that

$$\lim_{N\to\infty} E(F) = \lim_{N\to\infty} (N-2)E\left(\frac{R^2}{1-R^2}\right) = \infty \tag{39}$$

for the standard $F_{1,N-2}$ statistic on the significance of β_2. The non-existence of the probability limit of b_2 forces Granger and Newbold to use simulation. In spite of (38) they find spurious results in finite samples.

They also consider the case in which η_n and ζ_n are MA errors i.e.

$$\eta_n = \tilde{\eta}_n - \theta_1 \tilde{\eta}_{n-1}$$
$$\zeta_n = \tilde{\zeta}_n - \theta_2 \tilde{\zeta}_{n-1} \tag{40}$$

with $\tilde{\eta}_n, \tilde{\zeta}_n$ white noise. They note that if $\theta_2 \to 1$ the spurious nature of the regression is hard to detect. This is not surprising given the fact that if $\theta_2 = 1$ we have

$$y_n = \mu n + \tilde{\zeta}_n$$

i.e. model (34) satisfies (B.1)-(B.5). Except in this extreme case, the previous findings and in particular(35) also hold for error terms of the MA type.

The observation that the non-convergence of the OLS estimator b_2 is crucial to Granger and Newbold's results suggests a cross-section analogue to spurious regression. Consider two independent random variables y,x where y follows a Cauchy distribution. The Cauchy distribution has no moments and can be used to describe the frequency distribution of a variable with (a large number of) 'outlying' observations. The observations y_n, x_n are independent and we assume that x has a finite variance. If we use the simple regression model (34) to analyse the relation between y and x, we find that the OLS estimator b_2 has no probability limit. Again R^2 does not converge to 0 and (39) also holds in this case. The apparent relation is spurious. Note that model (34) violates assumption (B.2). In this case transforming the dependent variable y to a distribution with a finite mean e.g. a normal distribution would reveal the spurious nature of the relation.

7. Concluding remarks

From the foregoing discussion we conclude that validity of the (B)-type assumptions is essential for reliability and interpretability of regression results. Inconsistent coefficient estimation as in spurious regressions only occurs in situations where we use a model for which these assumptions are violated. Testing the disturbances for zero mean, homoscedasticity, serial independence, orthogonality with respect to the regressors, and perhaps for normality should be standard practice. By such analysis, which mainly concentrates on the regression residuals, the regression results may be either refuted or vindicated. Also transforming the model may be helpful for discovering and possibly for curing the spurious nature of estimated relations.

Zellner (1982, p. 88) objects in a footnote against the phrase 'ordinary least squares', since he does not think the least-squares principle to be ordinary. We think it to be highly desirable, that it will become ordinary (i.e. usual, customary, routine, normal) in the practice of regression analysis that OLS results are always supplemented with the above-mentioned validation checks of the fundamental regression model assumptions.

Appendix

Consider the sequence of OLS estimators for the slope coefficient in (34)

$$(A.1) \quad b_{2N} = \frac{\sum_{n=1}^{N} (x_n - \bar{x}_N)(y_n - \bar{y}_N)}{\sum_{n=1}^{N} (x_n - \bar{x}_N)^2} = \frac{\frac{1}{N}\sum_{n=1}^{N} x_n y_n - \bar{x}_N \bar{y}_N}{\frac{1}{N}\sum_{n=1}^{N} x_n^2 - \bar{x}_N^2}$$

where $\bar{x}_N, \bar{y}_N$ denote the sample averages of x, y for $n = 1, \ldots, N$. We have after some tedious algebra

$$E\left(\frac{1}{N^3}\sum_{n=1}^{N} x_n y_n\right) = \frac{\mu\lambda}{3} + O\left(\frac{1}{N}\right)$$

$$\text{Var}\left(\frac{1}{N^3}\sum_{n=1}^{N} x_n y_n\right) = O\left(\frac{1}{N}\right)$$

$$E\left(\frac{1}{N}\bar{x}_N\right) = \frac{\lambda}{2} + O\left(\frac{1}{N}\right)$$

$$\text{Var}\left(\frac{1}{N}\bar{x}_N\right) = O\left(\frac{1}{N}\right)$$

$$E\left(\frac{1}{N^3}\sum_{n=1}^{N} x_n^2\right) = \frac{\lambda^2}{3} + O\left(\frac{1}{N}\right)$$

$$\text{Var}\left(\frac{1}{N^3}\sum_{n=1}^{N} x_n^2\right) = O\left(\frac{1}{N^2}\right)$$

We prove the following lemma:

Lemma

Let $\{z_N\}$ be a sequence of random variables with $E(z_N) \to \mu$ and $\text{var}(z_N) \to 0$, then $z_N \to \mu$ in probability.

Proof

We have by Chebyshev's inequality

$$\Pr(|z_N - \mu| \geq \varepsilon) \leq \frac{\text{Var}(z_N) + (\mu_N - \mu)^2}{\varepsilon^2}.$$

Hence for all $\varepsilon > 0$

$$\lim_{N \to \infty} \Pr(|z_N - \mu| \geq \varepsilon) = 0.$$

By substituting for z_N the random variables $(\sum_{n=1}^{N} x_n y_n)/N^3$, $\bar{x}_N/N$, $\bar{y}_N/N$ and $(\sum_{n=1}^{N} x_n^2)/N^3$ and substituting the probability limits in (A.1) we find

$$\plim_{N \to \infty} b_{2N} = \frac{\mu}{\lambda}.$$

Define

$$\tilde{b}_{1N} = \bar{y}_N - \frac{\mu}{\lambda} \bar{x}_N.$$

Then

$$\tilde{b}_{1N} = \frac{1}{N} \sum_{n=1}^{N} \left(w_n - \frac{\mu}{\lambda} v_n \right)$$

with

$$w_n = \sum_{l=1}^{n} \zeta_l \text{ and } v_n = \sum_{l=1}^{n} \eta_l.$$

It follows that

(A.2) $E(\tilde{b}_{1N}) = 0$

(A.3) $\lim_{N \to \infty} \text{Var}(\tilde{b}_{1N}) = \infty.$

Consider for arbitrary c

$$\Pr(\tilde{b}_{1N} \geq c) = \Pr\left(\frac{\tilde{b}_{1N}}{\sqrt{\text{Var}(\tilde{b}_{1N})}} \geq \frac{c}{\sqrt{\text{Var}(\tilde{b}_{1N})}} \right).$$

The random variable on the right-hand side has mean 0 and variance 1. Because of (A.2) and (A.3) we find

$$\lim_{N \to \infty} \Pr(\tilde{b}_{1N} \geq c) \geq \varepsilon > 0$$

for some $\varepsilon > 0$. A similar argument gives that

$$\lim_{N \to \infty} \Pr(\tilde{b}_{1N} \leq c) \geq \varepsilon > 0$$

for arbitrary c.

Next consider the case $\mu = \lambda = \gamma = 0$. The slope coefficient (we fit a model without an intercept) then is given by

$$b_{2N} = \frac{\sum_{n=1}^{N} v_n w_n}{\sum_{n=1}^{N} v_n^2}.$$

Then

$$E(b_{2N}) = E_{V_1, \ldots, V_N}\left[E\left[\frac{\sum_{n=1}^{N} v_n w_n}{\sum_{n=1}^{N} v_n^2}\,\Big|\, v_1, \ldots, v_N\right]\right]$$

$$= E_{V_1, \ldots, V_N}\left[\frac{\sum_{n=1}^{N} v_n E(w_n)}{\sum_{n=1}^{N} v_n^2}\right] = 0.$$

Further we have for arbitrary $\varepsilon > 0$

$$\Pr\left(\frac{\sum_{n=1}^{N} v_n w_n}{\sum_{n=1}^{N} v_n^2} > \varepsilon\right) = \Pr\left(\sum_{n=1}^{N} (v_n w_n - \varepsilon v_n^2) > 0\right).$$

If

$$z_N = \sum_{n=1}^{N} (v_n w_n - \varepsilon v_n^2)$$

then

$$E(z_N) = -\varepsilon \frac{N(N+1)}{2} \sigma_\eta^2$$

$$\text{Var}(z_N) = h(N)$$

with

$$\lim_{N \to \infty} \frac{h(N)}{N^4} > 0.$$

Hence

$$\Pr(z_N > 0) = \Pr\left(\frac{z_N - E(z_N)}{\sqrt{\text{Var}(z_N)}} > \frac{\varepsilon \frac{N(N+1)}{2} \sigma_\eta^2}{\sqrt{h(N)}}\right)$$

and

$$\lim_{N \to \infty} \Pr(z_N > 0) = \lim_{N \to \infty} \Pr\left(\frac{z_N - E(z_N)}{\sqrt{\text{Var}(z_N)}} > C\varepsilon\right).$$

Because the random variable on the right-hand side has mean 0 and variance 1 there is an $\varepsilon>0$ such that

$$\lim_{N\to\infty} \Pr(b_{2N}>\varepsilon)>0.$$

Because the model has no intercept, the coefficient of determination is given by

$$R_N^2 = \frac{\sum_{n=1}^{N} \hat{y}_n^2}{\sum_{n=1}^{N} y_n^2} = \frac{\left(\sum_{n=1}^{N} x_n y_n\right)^2}{\left(\sum_{n=1}^{N} x_n^2\right)\left(\sum_{n=1}^{N} y_n^2\right)}.$$

A similar argument as for b_{2N} gives that there is an $\varepsilon>0$ such that

$$\lim_{N\to\infty} \Pr(R_N^2>\varepsilon)>0.$$

Hence $E(R_N^2)>0$ for N large enough.

Finally we consider the cross-section case. Let y_n be (standard) Cauchy; x_n can be treated as a constant. Then, b_{2N} has a Cauchy distribution with scale parameter

$$\delta_N = \sum_{n=1}^{N} \frac{|x_n - \bar{x}_N|}{\sum_{n=1}^{N} (x_n - \bar{x}_N)^2}.$$

Of course,

$$\delta_N \to \frac{E|x-\mu|}{\mathrm{Var}(x)} \quad \text{a.e.}$$

where $\mu = E(x)$ and we assume that $\mathrm{Var}(x)$ is finite. Hence, b_{2N} converges in distribution to this Cauchy distribution (the limiting characteristic function is continuous in 0).

Bibliography

Cramèr H. 1945. Mathematical methods of statistics (tenth printing 1963). Princeton: Princeton University Press.

Cramer JS. 1973a. Empirical Econometrics (first printing 1969). Amsterdam: North-Holland Publishing Co.

Cramer JS. 1973b. Interaction of income and price in consumer demand. International Economic Review 14: 351–363

Granger CWJ, Newbold P. 1974. Spurious regressions in econometrics. Journal of Econometrics 2: 111–120.

Granger CWJ, Newbold P. 1977. Forecasting Economic Time Series. London: Academic Press.

Griliches Z, Intriligator MD. 1983, 1984. Editors of the Handbook of Econometrics, I, II. Amsterdam: North-Holland Publishing Co.

Kiviet JF, Phillips GDA. 1986. Testing strategies for model specification. Applied mathematics and computation 20: 237–269.

Malinvaud E. 1964. Méthodes statistiques de l'Econométrie. Paris: Dunod.

Pagan AR. 1984. Model evaluation by variable addition. In: Econometrics and quantitative economics, edited by DF Hendry and KF Wallis. Oxford: Basil Blackwell.

Ruud PA. 1984. Tests of specification in econometrics (with discussion). Econometric Reviews 3: 211–276.

White H. 1980. Using least squares to approximate unknown regression functions. International Economic Review 21: 149–170.

Wold HOA. 1953. Demand analysis, a study in econometrics (fourth printing 1966). New York: John Wiley and Sons.

Zellner A. 1982. Basic issues in Econometrics. Chicago: The University of Chicago Press.

The classical econometric model

STEPHEN POLLOCK

1. Introduction

For a great many years, and certainly for as long as Cramer has held the chair of econometrics at the University of Amsterdam, the simultaneous-equation model has represented, to econometric theorists, what Kuhn, the author of a famous monograph on the Structure of Scientific Revolutions (1972) would describe as a central paradigm of their science.

The simultaneous-equation model, which we shall hereafter call the classical model, was the subject of the research of the Cowles Commission (1949, 1950, 1953) in the late forties and early fifties which inaugurated the era of modern econometrics. The same model was a major preoccupation of the Dutch econometricians in the late fifties and early sixties, and substantial contributions to its interpretation were made by Theil who propounded the techniques of two-stage and three-stage least-squares estimation (see Theil (1958) and Zellner and Theil (1962). Latterly, the model has been the focus of research into the statistical distributions of estimates derived from small samples; and it continues to stimulate researches into the theory of matrix differential calculus of the sort that have been pursued with particular vigour in Cramer's own department (see Neudecker (1969) and the bibliography in Pollock (1985)).

The characteristics of the classical simultaneous-equation model which inspire this enduring interest can be described in terms of a simile. The classical model is like a hard and dense crystalline object with many polished facets. To each of these facets there corresponds a point of view from which one can look into the object. From most points of view, what we are likely to see is a myriad of bewildering internal reflections. However, there are some axes or lines of sight which render the object transparent. It is the density of the model, as well as the fact that, if it is approached in an appropriate way, we can see through it, that has made it so attractive to the theorists.

In this paper, our object is to provide a compendium of the various points of view that may be taken in deriving the estimators of the parameters of the classical model according to the principle of maximum likelihood. The synoptic comparisons which this will enable should help in overcoming the bewilderment that can arise from seeing so many different forms of the estimating equations. We believe that the problems of estimation are best understood by concentrating primarily on three approaches which give rise, severally, to what we shall describe as the Π method, the Ω method and the λ method of estimation.

2. The classical model

The classical model describes a stochastic relationship between a set of K exogenous variables contained in the vector $x' = [x_1, \ldots, x_K]$ and a set of G endogenous variables contained in the vector $y' = [y_1, \ldots, y_G]$. A single realization of the model can be written as

$$y'\Gamma + x'B = \varepsilon', \qquad (1)$$

where $\varepsilon' = [\varepsilon_1, \ldots, \varepsilon_G]$ is a normally distributed disturbance vector which has an expected value of $E(\varepsilon) = 0$ and a dispersion matrix of $D(\varepsilon) = \Psi$. The vectors x and ε are assumed to be uncorrelated, so that their covariance matrix is $C(x,\varepsilon) = 0$.

The diagonal elements of the nonsingular parameter matrix Γ are assumed to be all equal to minus one, and this indicates that y_j is to be regarded as the dependent variable of the jth equation. If e_j is the vector with a unit in the jth position and zeros elsewhere, then these restrictions, which are called the normalization rules, can be written as $e_j'\Gamma e_j = -1; j = 1, \ldots, G$. In addition, we assume that certain of the elements of Γ and B are zeros. Thus, if $\gamma_j = \Gamma e_j$ and $\beta_j = Be_j$ are the parameter vectors of the jth equation, we can write the full set of restrictions affecting them as

$$\begin{bmatrix} R'_{\gamma j} & 0 \\ 0 & R'_{\beta j} \end{bmatrix} \begin{bmatrix} \gamma_j \\ \beta_j \end{bmatrix} = \begin{bmatrix} r_j \\ 0 \end{bmatrix} \text{ or } \begin{bmatrix} R'_{\gamma j} & 0 \\ 0 & R'_{\beta j} \end{bmatrix} \begin{bmatrix} \gamma_j + e_j \\ \beta_j \end{bmatrix} = \begin{bmatrix} 0 \\ 0 \end{bmatrix}, \qquad (2)$$

where $R_{\beta j}$ comprises a selection of columns from the identity matrix I_K of order K, $R_{\gamma j} = [e_j, H_{\gamma j}]$ comprises, likewise, a set of columns from the identity matrix I_G of order G, and r_j is a vector containing zeros and a leading element of minus one corresponding to the normalization rule. We can represent the general solution to these restrictions by

$$\begin{bmatrix} \gamma_j \\ \beta_j \end{bmatrix} = \begin{bmatrix} P_{\gamma j} & 0 \\ 0 & P_{\beta j} \end{bmatrix} \begin{bmatrix} \gamma_{\Delta j} \\ \beta_{\Delta j} \end{bmatrix} - \begin{bmatrix} e_j \\ 0 \end{bmatrix}, \qquad (3)$$

where $\gamma_{\Delta j}$ and $\beta_{\Delta j}$ are composed of the G_j and K_j unrestricted elements of γ_j and

β_j respectively, and where $P_{\gamma j}$ and $P_{\beta j}$ are the complements of $R_{\gamma j}$ and $R_{\beta j}$ within I_G and I_K respectively.

The reduced form of equation (1) is written as

$$\begin{aligned} y' &= -x'B\Gamma^{-1} + \varepsilon'\Gamma^{-1} \\ &= x'\Pi + \eta', \end{aligned} \qquad (4)$$

where $\Pi = -B\Gamma^{-1}$ and $\eta' = \varepsilon'\Gamma^{-1}$ with $D(\eta) = \Gamma'^{-1}\Psi\Gamma^{-1} = \Omega$.

We shall denote the dispersion matrices of the data vectors by $D(x) = \sum_{xx}$, $D(y) = \sum_{yy}$ and their covariance matrix by $C(x, y) = \sum_{xy}$, and we shall assume that all of these attain the maximum rank. Then, from our existing assumptions, it follows that

$$\begin{bmatrix} \Gamma' & B' \\ 0 & I \end{bmatrix} \begin{bmatrix} \sum_{yy} & \sum_{yx} \\ \sum_{xy} & \sum_{xx} \end{bmatrix} \begin{bmatrix} \Gamma & 0 \\ B & I \end{bmatrix} = \begin{bmatrix} \Psi & 0 \\ 0 & \sum_{xx} \end{bmatrix}, \qquad (5)$$

and from this we are able to obtain equivalent expressions in the form of

$$\begin{bmatrix} \sum_{yy} & \sum_{yx} \\ \sum_{xy} & \sum_{xx} \end{bmatrix} \begin{bmatrix} \Gamma & 0 \\ B & I \end{bmatrix} = \begin{bmatrix} \Gamma'^{-1} & \Pi' \\ 0 & I \end{bmatrix} \begin{bmatrix} \Psi & 0 \\ 0 & \sum_{xx} \end{bmatrix} - \begin{bmatrix} \Omega\Gamma & \Pi'\sum_{xx} \\ 0 & \sum_{xx} \end{bmatrix} \qquad (6)$$

and

$$\begin{bmatrix} \sum_{yy} & \sum_{yx} \\ \sum_{xy} & \sum_{xx} \end{bmatrix} = \begin{bmatrix} \Gamma'^{-1} & \Pi' \\ 0 & I \end{bmatrix} \begin{bmatrix} \Psi & 0 \\ 0 & \sum_{xx} \end{bmatrix} \begin{bmatrix} \Gamma^{-1} & 0 \\ \Pi & I \end{bmatrix} =$$

$$\begin{bmatrix} \Pi'\sum_{xx}\Pi + \Omega & \Pi'\sum_{xx} \\ \sum_{xx}\Pi & \sum_{xx} \end{bmatrix}. \qquad (7)$$

The above identities provide us with the fundamental equation that relate the structural parameters γ_j, β_j; $j = 1, \ldots,$ G to the moment matrices of the data variables. We can write these equations in two alternative forms:

$$\begin{bmatrix} 0 \\ 0 \end{bmatrix} = \begin{bmatrix} \sum_{yy} - \Omega & \sum_{yx} \\ \sum_{xy} & \sum_{xx} \end{bmatrix} \begin{bmatrix} \gamma_j \\ \beta_j \end{bmatrix} = \begin{bmatrix} \Pi'\sum_{xy} & \Pi'\sum_{xx} \\ \sum_{xy} & \sum_{xx} \end{bmatrix} \begin{bmatrix} \gamma_j \\ \beta_j \end{bmatrix}. \qquad (8)$$

The first of these follows directly from (6). The second is obtained via the identities $\sum_{yy} = \Pi'\sum_{xx}\Pi + \Omega$ and $\sum_{xy} = \sum_{xx}\Pi$ that are contained in (7).

It is clear, from the second expression under (8), that all the information relating to γ_j and β_j that is provided by that equation is contained in its constituent part $\sum_{xy}\gamma_j + \sum_{xx}\beta_j = 0$. On substituting the solutions $\gamma_j = P_{\gamma j}\gamma_{\Delta j} - e_j$ and $\beta_j = P_{\beta j}\beta_{\Delta j}$ from equation (3) into this expression, we get

$$\sum_{xy} P_{\gamma j}\gamma_{\Delta j} + \sum_{xx} P_{\beta j}\beta_{\Delta j} = \sum_{xy} e_j. \qquad (9)$$

This is a set of K equations in $G_j + K_j$ unknowns, and, given that the matrix $[\sum_{xy}, \sum_{xx}]$ is of full rank, it follows that the necessary and sufficient condition for the identifiability of the parameters vectors of the jth equation is that

$K \geq G_j + K_j$. If this condition is fulfilled, then any subset of $G_j + K_j$ of the equations of (9) will serve to determine $\gamma_{\Delta j}$ and $\beta_{\Delta j}$. However, we shall be particularly interested in a set of $G_j + K_j$ independent equations in the form of

$$\begin{bmatrix} P'_{yj}\Pi'\Sigma_{xy}P_{yj} & P'_{yj}\Pi'\Sigma_{xx}P_{\beta j} \\ P'_{\beta j}\Sigma_{xy}P_{yj} & P'_{\beta j}\Sigma_{xx}P_{\beta j} \end{bmatrix} \begin{bmatrix} \gamma_{\Delta j} \\ \beta_{\Delta j} \end{bmatrix} = \begin{bmatrix} P'_{yj}\Pi'\Sigma_{xy}e_j \\ P'_{\beta j}\Sigma_{xy}e_j \end{bmatrix} \quad (10)$$

which are derived by premultiplying equation (9) by the matrix $[\Pi P_{yj}, P_{\beta j}]'$. These equations, which we have derived solely by considering the relationship between the parameters of our model and the moments of the data vectors x and y, must be the basis of any reasonable estimator of the parameters of the individual structural equations, regardless of the principles from which it is derived.

3. The maximum-likelihood estimator of single equations

In describing the procedures by which we may make inferences about the parameters of the model, we shall assume that we have a set $T \geq G + K$ observations on the data vectors which we shall denote by $x_t, y_t; t = 1, \ldots, T$. From these observations, we may construct a set of moment matrices in the forms of $S_{xx} = \sum_t x_t x'_t / T$, $S_{xy} = \sum_t x_t y'_t / T$ and $S_{yy} = \sum_t y_t y'_t / T$ which we may assume to have full rank. These are the empirical counterparts of the moment matrices Σ_{xx}, Σ_{xy} and Σ_{yy} which characterize the distribution of the population.

The principle of estimation known as the method of moments suggests that we should obtain our estimating equations by replacing the population moments within our equations (7) and (9) by the empirical moments. Thus, according to this principle, equation (7) yields an estimate $S_{xx}^{-1} S_{xy}$ of the reduced-form parameter matrix Π which is simply the unrestricted least-squares estimator. Likewise, in the case where $G_j + K_j = K$, equation (9) yields the so-called indirect least-squares estimator of the structural parameters $\gamma_{\Delta j}$ and $\beta_{\Delta j}$.

However, this simple method of moments is not adequate in the case of the overidentified model where $G_j + K_j < K$. For then the restrictions on the vectors γ_j and β_j imply restrictions on Σ_{xy} and Σ_{yy}; and the effect is that, in the context of the set of all values that might be assumed by S_{xy} and S_{yy}, the subset containing the values that will render the equations in (9) algebraically consistent has a zero measure.

There are two alternative approaches that may be followed in the attempt to overcome the problem of overidentification. The first approach is to resolve the algebraic inconsistency of the equations

$$S_{xy} P_{yj} \gamma_{\Delta j} + S_{xx} P_{\beta j} \beta_{\Delta j} = S_{xy} e_j. \quad (11)$$

via a method of least-squares regression. Thus, by using a regression metric defined in terms of the matrix S_{xx}^{-1}, we may derive the two-stage least-squares estimator of Theil (1958) and Basmann (1957). What is rather remarkable is the fact that this is precisely the estimator which is also derived from the equation (10) when $\sum_{xy}$ and $\sum_{yy}$ are replaced by S_{xy} and S_{yy} respectively and Π is represented by its unrestricted ordinary least-squares estimator.

Another way of overcoming the problem of overidentification is to find a set of admissible and mutually conformable estimates of $\sum_{xy}$, $\sum_{yy}$, γ_j and β_j which satisfy equation (9) exactly. This is what the method of maximum likelihood achieves.

3.1 The Π method

There are a number of different yet equivalent criteria from which we may derive the maximum-likelihood estimators of the structural parameters. We shall begin, as the author has done in [13] and [14], by using, as our criterion function, an expression of the likelihood function which is in terms of the reduced-form parameters:

$$L(\Pi,\Omega) = (2\mu)^{-GT/2}|\Omega|^{-T/2}.$$
$$\exp\{(-T/2)\text{Trace}(S_{yy} - S_{yx}\Pi - \Pi'S_{xy} + \Pi'S_{xx}\Pi)\Omega^{-1}\}. \tag{12}$$

This is also the starting point for the classical derivation of Anderson and Rubin (1949).

In order to convey to the reduced-form parameters the effects of the structural-form restrictions under (2), we must use the condition $\Pi\gamma_j + \beta_j = 0$ which comes directly from the identity $\Pi = -B\Gamma^{-1}$ by which we have defined the reduced-form parameters. Thus the appropriate criterion function is given by the expression

$$C(\Pi,\Omega,\gamma_j,\beta_j) = \log L(\Pi,\Omega) - \varkappa'(\Pi\gamma_j + \beta_j) - \mu'(Q'_{\gamma j}\gamma_j + Q'_{\beta j}\beta_j), \tag{13}$$

where $Q_{\gamma j} = [R_{\gamma j}, 0]$, $Q_{\beta j} = [0, R_{\gamma j}]$, and where $\varkappa$ and λ are vectors of Lagrangean multipliers.

By differentiating the criterion function with respect to Π, Ω and $\varkappa$ and setting the results to zero for a maximum, we obtain the conditions from which we may derive our estimating equations for Π and Ω. The latter are respectively

$$\Pi = S_{xx}^{-1}S_{xy} - S_{xx}^{-1}(S_{xy}\gamma_j + S_{xx}\beta_j)(\gamma_j'W\gamma_j)^{-1}\gamma_j'W \tag{14}$$

and

$$\Omega = W - W\gamma_j(\gamma_j'W\gamma_j)^{-1}(\gamma_j'S_{yx} + \beta_j'S_{xx})S_{xx}^{-1}(S_{xy}\gamma_j + S_{xx}\beta_j).$$
$$(\gamma_j'W\gamma_j)^{-1}\gamma_j'W, \tag{15}$$

where $W = S_{yy} - S_{yx}S_{xx}^{-1}S_{xy}$ is the ordinary unrestricted estimate of $\Omega = \Sigma_{yy} - \Pi'\Sigma_{xx}\Pi$. It is straightforward to confirm that the restricted estimator of Π does satisfy the condition $\Pi\gamma_j + \beta_j = 0$.

Given that Σ_{xx} is always estimated by S_{xx}, regardless of the restrictions on γ_j and β_j, it follows that the restricted estimator of $\Sigma_{xy} = \Sigma_{xx}\Pi$ is given by $S_{xx}\Pi$, where Π stands for the expression in (14). When these estimates are used to represent Σ_{xy} and Σ_{xx} within the expression under (9), we get

$$S_{xx}\Pi P_{\gamma j}\gamma_{\Delta j} + S_{xx}P_{\beta j}\beta_{\Delta j} = S_{xx}\Pi e_j. \tag{16}$$

This expression represents nothing more than an algebraic identity. Therefore, unlike the equation (11) upon which the two-stage least-squares estimator is based, there is no question of any algebraic inconsistency when $K > K_j + G_j$.

To obtain the maximum-likelihood estimating equations for the structural parameters, we differentiate the criterion function in respect of γ_j and β_j and set the results to zero. Then, by substituting for $\varkappa = (S_{xy}\gamma_j + S_{xx}\beta_j)(\gamma_j'W\gamma_j)^{-1}$, we obtain the equation

$$\begin{bmatrix} \Pi'S_{xy} & \Pi'S_{xx} \\ S_{xy} & S_{xx} \end{bmatrix} \begin{bmatrix} \gamma_j \\ \beta_j \end{bmatrix} + \begin{bmatrix} R_{\gamma j} & 0 \\ 0 & R_{\beta j} \end{bmatrix} \begin{bmatrix} \varphi_\gamma \\ \varphi_\beta \end{bmatrix} = \begin{bmatrix} 0 \\ 0 \end{bmatrix}, \tag{17}$$

where $[\varphi_\gamma'\ \varphi_\beta']' = \varphi = \mu(\gamma_j'W\gamma_j)$. On substituting $\gamma_j = P_{\gamma j}\gamma_{\Delta j} - e_j$ and $\beta_j = P_{\beta j}\beta_{\Delta j}$ from equation (3) into this expression, and by premultiplying it by the transpose of the matrix in (3), we get

$$\begin{bmatrix} P_{\gamma j}'\Pi'S_{xy}P_{\gamma j} & P_{\gamma j}'\Pi'S_{xx}P_{\beta j} \\ P_{\beta j}'S_{xy}P_{\gamma j} & P_{\beta j}'S_{xx}P_{\beta j} \end{bmatrix} \begin{bmatrix} \gamma_{\Delta j} \\ \beta_{\Delta j} \end{bmatrix} = \begin{bmatrix} P_{\gamma j}'\Pi'S_{xy}e_j \\ P_{\beta j}'S_{xy}e_j \end{bmatrix}. \tag{18}$$

This is precisely what one derives from the equation (10) when Σ_{xx} and Σ_{xy} are replaced by S_{xx} and S_{xy} respectively and Π is represented by its restricted estimator from (14).

If we take the above equation together with equation (14) for the restricted estimator of Π, then we have a complete estimating system for $\gamma_j = P_{\gamma j}\gamma_{\Delta j} - e_j$ and $\beta_j = P_{\beta j}\beta_{\Delta j}$ which can be solved iteratively. That is to say, given an initial value Π_0, we can obtain first-round estimates of $\gamma_{\Delta j}$ and $\beta_{\Delta j}$ from equation (18). On taking these values to equation (14), we can obtain a revised estimate for Π for use in the second round; and we can extend this procedure through an indefinite number of iterations in the expectation, that, eventually, the estimates from successive rounds will become virtually identical. We shall describe this procedure as the Π method of estimation.

3.2 The Ω method

To obtain our second version of the estimating equations, we use the identity

$$(S_{yy} - \Omega)\gamma_j + S_{yx}\beta_j = \Pi' S_{xy}\gamma_j + \Pi' S_{xx}\beta_j \tag{19}$$

which holds when Ω and Π are represented by the expressions in (14) and (15). This identity enables us to rewrite the estimating equations for $\gamma_{\Delta j}$ and $\beta_{\Delta j}$ as

$$\begin{bmatrix} P'_{\gamma j}(S_{yy} - \Omega)P_{\gamma j} & P'_{\gamma j}S_{xy}P_{\beta j} \\ P'_{\beta j}S_{xy}P_{\gamma j} & P'_{\beta j}S_{xx}P_{\beta j} \end{bmatrix} \begin{bmatrix} \gamma_{\Delta j} \\ \beta_{\Delta j} \end{bmatrix} = \begin{bmatrix} P'_{\gamma j}(S_{yy} - \Omega)e_j \\ P'_{\beta j}S_{xy}e_j \end{bmatrix}. \tag{20}$$

This form corresponds to the first of the two equations under (8) which describe the relationship between the moments of the data variables and the parameters of the jth equation. If we take these equations (20) together with our equation (15) for the restricted estimate of Ω, then we have an alternative estimating system that can be solved iteratively for Ω, γ_j and β_j. Thus we have what we shall call the Ω method of estimation.

3.3 The λ method

To obtain our third estimating system, we use an identity which we can write as

$$\Omega \gamma_j = \lambda W \gamma_j, \tag{21}$$

where W is our unrestricted estimator of Ω and

$$\lambda = (\gamma'_j S_{yy} \gamma_j + \gamma'_j S_{yx} \beta_j + \beta'_j S_{xy} \gamma_j + \beta'_j S_{xx} \beta_j)(\gamma'_j W \gamma_j)^{-1}. \tag{22}$$

On substituting the identity into (20), the estimating equations for $\gamma_{\Delta j}$ and $\beta_{\Delta j}$ become

$$\begin{bmatrix} P'_{\gamma j}(S_{yy} - \lambda W)P_{\gamma j} & P'_{\gamma j}S_{xy}P_{\beta j} \\ P'_{\beta j}S_{xy}P_{\gamma j} & P'_{\beta j}S_{xx}P_{\beta j} \end{bmatrix} \begin{bmatrix} \gamma_{\Delta j} \\ \beta_{\Delta j} \end{bmatrix} = \begin{bmatrix} P'_{\gamma j}(S_{yy} - \lambda W)e_j \\ P'_{\beta j}S_{xy}e_j \end{bmatrix}. \tag{23}$$

Our sequence of estimates of γ_j and β_j is now obtained by solving the equations (22) and (23) in successive rounds. This constitutes the λ method of estimation.

The equations of the λ method come closest to the classical limited-information maximum-likelihood estimating equations that are common to the expositions of Anderson and Rubin (1949) and Koopmans and Hood (1953). The outstanding difference is that the latter equations are written in a homogeneous form which results from the suppression of the normalization rule $\gamma_j e_j = -1$. In consequence of this difference, the factor λ assumes the role of a latent root which can be extracted by the power method.

When λ assumes the role of a latent root, an alternative interpretation of the structural equation is engendered. For then the estimating equations come to resemble those of an errors-in-variables model. In this context, it is appropriate to seek to estimate γ_j and β_j by minimizing the function λ subject to the restrictions in (3). This approach to the derivation of the estimating equations, which may, of course, be subsumed under the maximum-likelihood principle, is what Koopmans and Hood have described as the minimum variance-ratio method.

3.4 Computations

In order to specify completely any of the foregoing methods of estimation, we need to choose the initial conditions with which to begin the various iterative procedures. The natural choices for the initial values of Π and Ω are their unrestricted estimates that are given by $S_{xx}^{-1}S_{xy}$ and $S_{yy} - S_{yx}S_{xx}^{-1}S_{xy}$ respectively. The natural choice for a starting value for λ is unity, since this is the value to which it must tend asymptotically as the sample moments tend to the values of the population moments. It is remarkable that, with these choices, the first rounds of the Π, Ω and λ methods of estimation give rise to identical estimates of γ_j and β_j which are simply the two-stage least-squares estimates of Theil (1958) and Basmann (1957).

4. The maximum-likelihood estimation of subsystems

One of the more surprising aspects of the development of the theory of the classical model has been the lack, until recently, of a complete treatment of the problem of estimating subsystems of the model which comprise more than one equation. This state of affairs prevailed in spite of the fact that, in their seminal account of the single-equation estimator, Koopmans and Hood (1953) had provided a criterion function from which the estimating equations of a multi-equation subsystem may be derived directly.

The advantage of the subsystem estimator is that it can be specialized easily in either direction to provide, on the one hand, the estimator of a single equation and, on the other hand, the estimator of the system as a whole. Thus we can articulate a full theory of estimation by considering only the case of subsystems.

4.1 The Π method for subsystems

It transpires that the procedure which we have used in deriving the estimating equations for the Π method can be very easily generalized to cope with subsystems. The only complication is that, in order to derive the estimating equations of multi-equation subsystems, we have to deal in terms of the vectorization operator and the algebra of Kronecker products. An account of this algebra has been given by the author in (12).

Let us assume that the subsystem of interest has a total of $G_1 = G - G_2$ equations whose parameters are contained in the matrices $\Gamma_1 = \Gamma J_1$ and $B_1 =$

BJ_1, where J_1 is a matrix consisting of G_1 columns selected from the identity matrix of order G. Then the restrictions on these parameters, including the normalization rules, may be written as

$$R' \begin{bmatrix} \Gamma_1 + J_1 \\ B_1 \end{bmatrix}^c = 0 \text{ and } (\Gamma_1' \otimes I)\Sigma_{xy}^c + (B_1' \otimes I)S_{xx}^c, \tag{24}$$

where the superscript c denotes the vectorization operator. We can procede to form a Lagrangean expression which is a straightforward generalization of the expression under (13). Then, from the first-order conditions for the optimization of the Lagrangean, we can obtain the following estimators of Π and Ω:

$$\Pi = S_{xx}^{-1}S_{xy} - S_{xx}^{-1}(S_{xy}\Gamma_1 + S_{xx}B_1)(\Gamma_1'W\Gamma_1)^{-1}\Gamma_1'W, \tag{25}$$

$$\Omega = W - W\Gamma_1(\Gamma_1'W\Gamma_1)^{-1}(\Gamma_1'S_{yx} + B_1'S_{xx})S_{xx}^{-1}(S_{xy}\Gamma_1 + S_{xx}B_1).$$

$$(\Gamma_1'W\Gamma_1)^{-1}\Gamma_1'W. \tag{26}$$

These are the obvious generalizations of the expressions under (14) and (15). The estimating equations for Γ_1 and B_1, which are obtained in the same manner as in the single-equation case, are given by

$$\left\{ \begin{bmatrix} \Pi'S_{xy} & \Pi'S_{xx} \\ S_{xy} & S_{xx} \end{bmatrix} \begin{bmatrix} \Gamma_1 \\ B_1 \end{bmatrix} (\Gamma_1'\Omega\Gamma_1)^{-1} \right\}^c + R\mu = 0, \tag{27}$$

where μ is a vector of Lagrangean multipliers associated with the restrictions on Γ_1 and B_1. This is just a generalization of equation (17). Notice, however, that, whereas, in the single-equation case, we were able to absorb the scalar factor $\gamma_j'\Omega\gamma_j$ in the multiplier γ, in the multi-equation case, we have to represent the corresponding matrix factor $\Gamma_1'\Omega\Gamma_1$ explicitly.

To find a compact form of the estimating equations to compare with those under (18), we solve the first set of restrictions under (24) to obtain

$$\begin{bmatrix} \Gamma_1 \\ B_1 \end{bmatrix}^c = PP' \begin{bmatrix} \Gamma_1 \\ B_1 \end{bmatrix}^c - \begin{bmatrix} J_1 \\ 0 \end{bmatrix}^c, \tag{28}$$

where P is the complement of R within the identity matrix of order $G_1(G + K)$ and $[\Gamma_1', B_1']^{c'}P$ is a vector of the unrestricted elements of Γ_1 and B_1. On substituting this solution into (27) and on premultiplying that equation by P', we get

$$\left(P' \left\{ (\Gamma_1'\Omega\Gamma_1)^{-1} \otimes \begin{bmatrix} \Pi'S_{xy} & \Pi'S_{xx} \\ S_{xy} & S_{xx} \end{bmatrix} \right\} P \right) P' \begin{bmatrix} \Gamma_1 \\ B_1 \end{bmatrix}^c$$

$$= \left(P' \left\{ (\Gamma_1'\Omega\Gamma_1)^{-1} \otimes \begin{bmatrix} \Pi'S_{xy} & \Pi'S_{xx} \\ S_{xy} & S_{xx} \end{bmatrix} \right\} P \right) \begin{bmatrix} J_1 \\ 0 \end{bmatrix}^c. \tag{29}$$

4.2 The estimating equations of the Ω method

To derive our estimating equations of the Ω method directly, we may begin by reconsidering the likelihood function that is given under (12). The exponent in this expression can be written as

$$(S_{yy} - \Pi'S_{xy} - S_{yx}\Pi - \Pi'S_{xx}\Pi)\Omega^{-1} = (S_{yy} - S_{yx}S_{xx}^{-1}S_{xy})\Omega^{-1} +$$
$$(S_{yx}S_{xx}^{-1}S_{xy} - \Pi'S_{xy} - S_{yx}\Pi + \Pi'S_{xx}\Pi)\Omega^{-1}. \tag{30}$$

Using the $\Sigma_{xy} = S_{xx}\Pi$ to denote the restricted estimator Σ_{xy}, we can write the second term on the RHS as

$$(S_{yx}S_{xx}^{-1}S_{xy} - \Pi'S_{xy} - S_{yx}\Pi + \Pi'S_{xx}\Pi)\Omega^{-1}$$
$$= (S_{yx} - \Sigma_{yx})S_{xx}^{-1}(S_{xy} - \Sigma_{xy})\Omega^{-1}. \tag{31}$$

If our object is to find an estimator of Σ_{xy} that is admissible with respect to the various restrictions on Γ_1 and B_1, then it is appropriate to find the value which minimizes the following Lagrangean criterion function:

$$C(\Gamma_1, B_1, \Omega) = \frac{1}{2}(S_{xy} - \Sigma_{xy})^{c'}(\Omega \otimes S_{xx})^{-1}(S_{xy} - \Sigma_{xy})^c$$
$$+ \Phi^{c'}\{(\Gamma_1' \otimes I)\Sigma_{xy}^c + (B_1' \otimes I)S_{xx}^c\}. \tag{32}$$

By differentiating the function with respect to Σ_{xy}^c and setting the result to zero, we obtain the condition

$$(S_{xy} - \Sigma_{xy})^{c'}(\Omega \otimes S_{xx})^{-1} - \Phi^{c'}(\Gamma_1' \otimes I) = 0. \tag{33}$$

This gives us

$$(S_{xy} - \Sigma_{xy})^{c'}(\Omega \otimes S_{xx})^{-1}(S_{xy} - \Sigma_{xy})^c = \Phi^{c'}(\Gamma_1'\Omega\Gamma_1 \otimes S_{xx})\Phi^c, \tag{34}$$

as well as

$$\Phi^{c'} = (S_{xy} - \Sigma_{xy})^{c'}(\Gamma_1 \otimes I)(\Gamma_1'\Omega\Gamma_1 \otimes S_{xx})^{-1}. \tag{35}$$

However, using the condition, $\Sigma_{xy}\Gamma_1 + S_{xx}B_1 = 0$, we can rewrite the latter as

$$\Phi^{c'} = (S_{xy}\Gamma_1 + S_{xx}B_1)^{c'}(\Gamma_1'\Omega\Gamma_1 \otimes S_{xx})^{-1}. \tag{36}$$

On substituting this into (33) and rearranging the result, we get,

$$\Sigma_{xy} = S_{xy} - (S_{xy}\Gamma_1 + S_{xx}B_1)(\Gamma_1'\Omega\Gamma_1)^{-1}\Gamma_1'\Omega. \tag{37}$$

This is our restricted estimator of Σ_{xy} and, of course, in the form of $\Pi = S_{xx}^{-1}\Sigma_{xy}$, it provides us again with the restricted estimator of Π.

On substituting the expression from (36) into (34), we get

$$(S_{xy} - \Sigma_{xy})^{c'}(\Omega \otimes S_{xx})^{-1}(S_{xy} - \Sigma_{xy})^c$$

$$= (S_{yx}\Gamma_1 + S_{xx}B_1)^{c'} (\Gamma_1'\Omega\Gamma_1 \otimes S_{xx})^{-1} (S_{xy}\Gamma_1 + S_{xx}B_1)^c. \tag{38}$$

The function on the RHS is essentially a generalization of the criterion function used by Keller (1975) in his derivation of the single-equation estimator. To derive the criterion for estimating Γ_1 and B_1, we subject the function to the restrictions in (24). By differentiating the resulting Langrangean expression with respect to Γ_1 and B_1 and setting the results to zero, we obtain the condition

$$\left\{ \begin{bmatrix} S_{yx}S_{xx}^{-1}S_{xy} & S_{yx} \\ S_{xy} & S_{xx} \end{bmatrix} \begin{bmatrix} \Gamma_1 \\ B_1 \end{bmatrix} (\Gamma_1'\Omega\Gamma_1)^{-1} + \begin{bmatrix} Q \\ 0 \end{bmatrix} \right\}^c + R\mu = 0, \tag{39}$$

wherein

$$Q = \Omega\Gamma_1(\Gamma_1'\Omega\Gamma_1)^{-1}(\Gamma_1'S_{yx} + B_1'S_{xx})S_{xx}^{-1}(S_{xy}\Gamma_1 + S_{xx}B_1)(\Gamma_1'\Omega\Gamma_1)^{-1}$$
$$= (W\Gamma_1 - \Omega\Gamma_1)(\Gamma_1'\Omega\Gamma_1)^{-1}. \tag{40}$$

Here the second equality follows from equation (26) via the identity $W\Gamma_1(\Gamma_1'W\Gamma_1)^{-1} = \Omega\Gamma_1(\Gamma_1'\Omega\Gamma_1)^{-1}$. On substituting our expression for Q back into (39) and using $W = S_{yy} - S_{yx}S_{xx}^{-1}S_{xy}$, we get our estimating equation of the Ω method:

$$\left\{ \begin{bmatrix} S_{yy} - \Omega & S_{yx} \\ S_{xy} & S_{xx} \end{bmatrix} \begin{bmatrix} \Gamma_1 \\ B_1 \end{bmatrix} (\Gamma_1'\Omega\Gamma_1)^{-1} \right\}^c + R\mu = 0, \tag{41}$$

4.3 The λ method for subsystems

As in the cases of the Π method and the Ω method, the estimating equations of the λ method can be derived directly from their own version of the maximum-likelihood criterion function. This is the concentrated likelihood function of Koopmans and Hood (1953). The process by which these authors derived their concentrated function is lengthy and difficult, and we shall give only a brief summary of it.

We may begin by considering the structural form of the likelihood function:

$$L(\Gamma, B\ \Psi) = (2\mu)^{-GT/2}|\Psi|^{-T/2}|\Gamma|^T.$$
$$\exp\{(-T/2)\text{Trace}(\Gamma'S_{yy}\Gamma + \Gamma'S_{yx}B + B'S_{xy}\Gamma + B'S_{xx}B)\Psi^{-1}\}. \tag{42}$$

This is obtained from the reduced form of the likelihood function under (12) via the identities $\Pi = -B\Gamma^{-1}$ and $\Omega^{-1} = \Gamma'\Psi^{-1}\Gamma$. The matrices of the structural parameters are partitioned as

$$\begin{bmatrix} \Gamma \\ B \end{bmatrix} = \begin{bmatrix} \Gamma_1 & \Gamma_2 \\ B_1 & B_2 \end{bmatrix} \text{ and } \Psi = \begin{bmatrix} \Psi_{11} & \Psi_{12} \\ \Psi_{21} & \Psi_{22} \end{bmatrix}. \tag{43}$$

In addition to the parameters Γ_1, B_1 and Ψ_{11} of the subsystem in which we are interested, we are now encumbered by the parameters of a complementary subsystem in which we have no direct interest. These additional parameters must either have acceptable arbitrary values assigned to them, or else they must be eliminated somehow from the likelihood function. We begin by attributing the following form to the structural dispersion matrix:

$$\Psi = \begin{bmatrix} \Psi_{11} & 0 \\ 0 & I \end{bmatrix}; \qquad (44)$$

for, as Koopmans and Hood show, this choice will not affect the estimates of the sought-after parameters provided that the complementary subsystem is not rendered overidentified. With this specification, our likelihood function becomes

$$L(\Gamma, B, \Psi) = (2\mu)^{-GT/2} |\Psi_{11}|^{-T/2} |\Gamma|^T \cdot$$
$$\exp\{(-T/2)\text{Trace}(\Gamma_1' S_{yy} \Gamma_1 + \Gamma_1' S_{yx} B_1 + B_1' S_{xy} \Gamma_1 + B_1' S_{xx} B_1) \Psi_{11}^{-1}$$
$$- (T/2)\text{Trace}(\Gamma_2' S_{yy} \Gamma_2 + \Gamma_2' S_{yx} B_2 + B_2' S_{xy} \Gamma_2 + B_2' S_{xx} B_2)\}. \qquad (45)$$

This device does not wholly succeed in separating the parameters of the two subsystems as we might desire, since we cannot resolve the determinant $|\Gamma| = |\Gamma_1 \ \Gamma_2|$ into the product of separate functions of Γ_1 and Γ_2. Therefore we have to eliminate the remaining parameters of the second subsystem by partially maximizing L, or its log, with respect to Γ_2 and B_2. It can be shown that, at the maximum, we have

$$\Gamma_2' S_{yy} \Gamma_2 + \Gamma_2' S_{yx} B_2 + B_2' S_{xy} \Gamma_2 + B_2' S_{xx} B_2 = I \quad \text{and}$$
$$|\Gamma_1 \ \Gamma_2|^T = |\Gamma_1' W \Gamma_1|^{T/2} |W|^{-T/2}. \qquad (46)$$

On substituting these values into (45), we obtain a concentrated likelihood function in the form of

$$L(\Gamma_1, B_1, \Sigma_1) = (2\mu)^{-GT/2} |\Psi_{11}|^{-T/2} |\Gamma_1' W \Gamma_1|^{T/2} |W|^{-T/2} \cdot$$
$$\exp\{(-T/2)\text{Trace}(\Gamma_1' S_{yy} \Gamma_1 + \Gamma_1' S_{yx} B_1 + B_1' S_{xy} \Gamma_1 +$$
$$B_1' S_{xx} B_1) \Psi_{11}^{-1} - G_2 T/2\}. \qquad (47)$$

On differentiating the latter function with respect to Ψ_{11} and setting the result to zero, we obtain a condition which provides the estimating equation

$$\Psi_{11} = \Gamma_1' S_{yy} \Gamma_1 + \Gamma_1' S_{yx} B_1 + B_1' S_{xy} \Gamma_1 + B_1' S_{xx} B_1. \qquad (48)$$

A further concentration of the likelihood function is available if we substitute from (48) back into (47) to get

$$L(\Gamma_1, B_1) =$$
$$(2\mu)^{-GT/2}\{|\Gamma_1'W\Gamma_1|/|\Gamma_1'S_{yy}\Gamma_1 + \Gamma_1'S_{yx}B_1 + B_1'S_{xy}\Gamma_1 + B_1'S_{xx}B_1|\}^{T/2}$$
$$\cdot |W|^{-T/2}\exp\{-GT/2\} \tag{49}$$

Embedded in this concentrated criterion function is a ratio of determinants which is a generalization of the ratio λ of (22) whose minimization leads to a single-equation version of the estimating equations.

To derive the estimating equations of a multi-equation subsystem, we may subject the log of the concentrated likelihood function under (47) to the restrictions on Γ_1 and B_1. Then by evaluating the first-order conditions for the optimization of the resulting Langrangean function, we obtain the following estimating equations:

$$\left\{\begin{bmatrix} S_{yy} & S_{yx} \\ S_{xy} & S_{xx} \end{bmatrix}\begin{bmatrix} \Gamma_1 \\ B_1 \end{bmatrix}\Psi_{11}^{-1} - \begin{bmatrix} W & 0 \\ 0 & 0 \end{bmatrix}\begin{bmatrix} \Gamma_1 \\ B_1 \end{bmatrix}(\Gamma_1'W\Gamma_1)^{-1}\right\}^c + R\mu = 0. \tag{50}$$

At first sight, these do not appear to resemble very closely the equations of the Ω method given under (41). However, the equivalence of the two sets of equations is readily established via the identities $W\Gamma_1(\Gamma_1'W\Gamma_1)^{-1} = \Omega\Gamma_1 \cdot (\Gamma_1'\Omega\Gamma_1)^{-1}$ and $\Psi_{11} = \Gamma_1'\Omega\Gamma_1$.

The derivation of Koopmans and Hood owes its difficulty largely to their use of the structural form of the likelihood function in place of the reduced form. Let us take the latter function instead, with a given value of Ω, and let us also use the identity $\Psi = \Gamma'\Omega\Gamma$ to write the structural dispersion matrix in the form of

$$\Gamma'\Omega\Gamma = \begin{bmatrix} \Gamma_1'\Omega\Gamma_1 & 0 \\ 0 & I \end{bmatrix}. \tag{51}$$

Then the likelihood function, which is conditional on Ω, can be written as

$$L(\Gamma, B|\Omega) = (2\mu)^{-GT/2}|\Omega|^{-T/2} \times$$
$$\exp\{(-T/2)\text{Trace}(\Gamma_1'S_{yy}\Gamma_1 + \Gamma_1'S_{yx}B_1 + B_1'S_{xy}\Gamma_1 + B_1'S_{xx}B_1)(\Gamma_1'\Omega\Gamma_1)^{-1}$$
$$- (T/2)\text{Trace}(\Gamma_2'S_{yy}\Gamma_2 + \Gamma_2'S_{yx}B_2 + B_2'S_{xy}\Gamma_2 + B_2'S_{xx}B_2)\}. \tag{52}$$

It is clear that the estimates of Γ_1 and B_1 are to be found by minimizing the first of the trace expressions without reference to other parameters. In this way, we derive a set of estimating equations in the form of

$$\left\{\begin{bmatrix} S_{yy} & S_{yx} \\ S_{xy} & S_{xx} \end{bmatrix}\begin{bmatrix} \Gamma_1 \\ B_1 \end{bmatrix}(\Gamma_1'\Omega\Gamma_1)^{-1} - \begin{bmatrix} Q \\ 0 \end{bmatrix}\right\}^c + R\mu = 0, \tag{53}$$

where

$$Q = \Omega\Gamma_1(\Gamma_1'\Omega\Gamma_1)^{-1}(\Gamma_1'S_{yy}\Gamma_1 + \Gamma_1'S_{yx}B_1 + B_1'S_{xy}\Gamma_1 + B_1'S_{xx}B_1)(\Gamma_1'\Omega\Gamma_1)^{-1}. \tag{54}$$

If Ω is given by the expression in (26), then the identity

$$\Gamma_1'\Omega\Gamma_1 = \Gamma_1'S_{yy}\Gamma_1 + \Gamma_1'S_{yx}B_1 + B_1'S_{xy}\Gamma_1 + B_1'S_{xx}B_1 \tag{55}$$

prevails, and we obtain exactly the equations of the Ω method given under (41).

5. The estimation of the system as a whole

The virtue of the three derivations that we have provided in the previous section is that the resulting estimating procedures are applicable to subsystems of all sizes ranging from a single structural equation to the system as a whole. Therefore our account of the useful methods could well be concluded at this point. Nevertheless, it may be interesting to consider the forms of some other criterion functions that are applicable only to the case of the system as a whole.

The estimating approach of Koopmans, Rubin and Leipnik (1950), which, for a considerable time, was regarded as the definitive approach to systemwide estimation, is based directly on the structural form of the likelihood function given under (42). The derivation of the estimating equations can be simplified by concentrating this function by substituting for Ψ its maximum-likelihood estimating equation which is

$$\Psi = \Gamma'S_{yy}\Gamma + \Gamma'S_{yx}B + B'S_{xy}\Gamma + B'S_{xx}B. \tag{56}$$

Then the concentrated likelihood function takes the form of

$$L(\Gamma,B) = (2\mu)^{-GT/2}|\Gamma'S_{yy}\Gamma + \Gamma'S_{yx}B + B'S_{xy}\Gamma + B'S_{xx}B|^{-T/2}|\Gamma|^T \exp(-GT/2\}. \tag{57}$$

The basic criterion is therefore to minimize the function

$$|\Gamma'S_{yy}\Gamma + \Gamma'S_{yx}B + B'S_{xy}\Gamma + B'S_{xx}B|/|\Gamma'\Gamma| \tag{58}$$

subject to the restrictions $R'[\Gamma' - I, B']'^c = 0$ affecting Γ and B. This gives rise to a set of estimating equations in the form of

$$\left\{ \begin{bmatrix} \Gamma^{-1'} \\ 0 \end{bmatrix} - \begin{bmatrix} S_{yy} & S_{yx} \\ S_{xy} & S_{xx} \end{bmatrix} \begin{bmatrix} \Gamma \\ B \end{bmatrix} \Psi^{-1} \right\}^c - R\mu = 0. \tag{59}$$

The explicit use of the inverse of Γ in these equations means that they cannot be used as a model for the estimating equations of a subsystem.

A variant of the criterion in (58) is to minimize the function

$$|\Gamma'S_{yy}\Gamma + \Gamma'S_{yx}B + B'S_{xy}\Gamma + B'S_{xx}B|/|\Gamma'S_{yy}\Gamma|. \tag{60}$$

The inclusion of the factor S_{yy} in the denominator has no effect upon the minimizing values of Γ and B. Nevertheless, the resulting estimating equations take the altered form of

$$\left\{ \begin{bmatrix} S_{yy} & S_{yx} \\ S_{xy} & S_{xx} \end{bmatrix} \begin{bmatrix} \Gamma \\ B \end{bmatrix} \Psi^{-1} - \begin{bmatrix} S_{yy} & 0 \\ 0 & 0 \end{bmatrix} \begin{bmatrix} \Gamma \\ B \end{bmatrix} (\Gamma' S_{yy} \Gamma)^{-1} \right\}^c + R\mu = 0. \quad (61)$$

These equations, which were originally derived by Chow (1964) are the basis of a common textbook presentation (See Dhrymes (1970) and Klein (1974) inter alia). It is easy to see that they are equivalent to the previous equations under (59).

The final variant of the criterion function that should be considered is one that has been adopted by Scharf (1976) in deriving what he has described as the full-information K-matrix class estimator. This function is given by

$$|\Gamma' S_{yy} \Gamma + \Gamma' S_{yx} B + B' S_{xy} \Gamma + B' S_{xx} B|/|\Gamma' W \Gamma|, \quad (62)$$

wherein $W = S_{yy} - S_{yx} S_{xx}^{-1} S_{xy}$ is the unrestricted estimate of Ω. The estimating equations that are derivable from this criterion function are nothing but the full-information version of the estimating equations of our λ method. It helps in understanding this result to recognize that the function in (62) is a generalization of the function in (22) from which, we have asserted, one can derive the single-equation version of the estimating equations.

Bibliography

Anderson TW, Rubin H. 1949. Estimation of the parameters of a single equation in a complete system of stochastic equations. Annals of Mathematical Statistics, 20: 46–63.

Basmann RL. 1957. A generalized classical method of linear estimation of coefficients in a structural equation. Econometrica, 24: 77–83.

Chow GC. 1964. A comparison of alternative estimators for simultaneous equations. Econometrica, 32: 532–553.

Chow GC. 1968. Two methods of computing full-information maximum-likelihood estimates in simultaneous stochastic equations. International Economic Review, 9: 100–112.

Dhrymes PJ. 1970. Econometrics: statistical foundations and applications. New York: Harper and Row.

Keller WJ. 1975. A new class of limited-information estimators for simultaneous-equation systems. Journal of Econometrics, 3: 71–92.

Klein LJ. 1974. A textbook of econometrics (second edition). Englewood Cliffs: Prentice-Hall.

Koopmans TC, Rubin H, Leipnik RB. 1950. Measuring the equation systems of dynamic economics, chapter 2 in: Statistical inference in dynamic economic models, edited by T.C. Koopmans. Cowles Foundation for Research in Economics, Monograph 10, New York: John Wiley and Sons.

Koopmans TC, Hood WC. 1953. The estimation of simultaneous linear economic relationships, chapter 6 in: Studies in econometric method, edited by W.C. Hood and T.C. Koopmans. Cowles Foundation for Research in Economics, Monograph 14, New York: John Wiley and Sons.

Kuhn TS. 1972. The structure of scientific revolutions (second edition). Chicago: University of Chicago Press.

Neudecker H. 1969. Some theorems on matrix differentiation with special reference to Kronecker matrix products. Journal of the American Statistical Association, 64: 953–963.

Pollock DSG. 1979. The algebra of econometrics. Chichester: John Wiley and Sons.

Pollock DSG. 1983. Varieties of the LIML estimator. Australian Economic Papers, December: 499–506.

Pollock DSG. 1984. Two reduced-form approaches to the derivation of the maximum-likelihood estimators for simultaneous-equation systems. Journal of Econometrics, 24: 331–347.

Pollock DSG. 1985. Tensor products and matrix differential calculus. Linear Algebra and its Applications, 67: 169–193.

Scharf W. 1976. K-matrix-class estimators and the full-information maximum-likelihood estimator as a special case. Journal of Econometrics, 4: 41–45.

Theil H. 1958. Economic forecasts and economic policy. Amsterdam: North-Holland Publishing Co.

Zellner A, Theil H. 1962. Three-stage least squares: simultaneous estimation of simultaneous equations. Econometrica, 30: 54–78.

Bibliography of Jan Salomon Cramer

1955
(1) De conjunctuurtest. *De Economist*, 103: 737–750

1956
(1) Loonverschillen tussen industriele bedrijfstakken, 1920;–1939 en 1947–1953. *Economisch-Statistische Berichten*, 41: 31–33.

1957
(1) A dynamic approach to the theory of consumer demand. *Review of Economic Studies*, 24: 73–86

1958
(1) Ownership elasticities of durable consumer goods. *Review of Economic Studies*, 25: 87–96
(2) The depreciation and mortality of motorcars. *Journal of the Royal Statistical Society*, A, 121: 18–59

1959
(1) Private motoring and the demand for petrol. *Journal of the Royal Statistical Society*, A, 122: 334–347

1962
(1) *The ownership of major consumer durables.* Cambridge: Cambridge University Press
(2) Dépenses et revenus des ménages d'après l'enquête de 1956. *Consommation*, 4: 5–36
(3) Het onderzoek van het consumentengedrag. *De Economist*, 110: 228–244

1964
(1) Efficient grouping, regression and correlation in Engel curve analysis. *Journal of the American Statistical Association*, 59: 233–250

1965
(1) Huur en inkomen van Amsterdamse gezinnen in 1923, 1934, 1951 en 1959 (with B.M.S. van Praag). *De Economist*, 113: 169–189
(2) Estimation in medium-term econometric models: the experts' practice (with L.B.M. Mennes). In *Modelli econometrici per la programmazione*, Scuola di Statistica, Firenze: 351–358

1966
(1) Een prijsindex voor nieuwe personenauto's, 1950–1965. *Statistica Neerlandica*, 20: 215–224
(2) Levensduur en sloop van personenauto's in Nederland, 1950–1964. *Statistica Neerlandica*, 20: 225–240
(3) Une analyse de budget de famille par composantes principales. *Economie Appliquée*, 19: 250–268

1968
(1) Autoverkopen en autopark in Nederland, 1950–1970. *Statistica Neerlandica*, 22: 119–132
(2) Woonruimte, huren en woonwensen van Amsterdamse gezinnen in 1962. *De Economist*, 116: 198–215

1969
(1) *Empirical Econometrics*. Amsterdam: North-Holland Publishing Co.
(2) Meervoudige aankopen (with R.D.H. Heijmans). *Orbis Economicus*, 13: 20–30

1970
(1) Interaction of price and income in consumer demand. *European Economic Review*, 1: 428–435

1973
(1) Het houderschap van liquiditeiten in Nederland (with G.M. Reekers). Amsterdam: NIBE
(2) Interaction of income and price in consumer demand. *International Economic Review*, 14: 351–363

1974

(1) A hedonic price index for the Dutch car market (with Nellie Kroonenberg). *De Economist*, 122: 359–366.

1976

(1) Money demand by sector (with G.M. Reekers). *Journal of Monetary Economics*, 2: 99–112
(2) The effect of income redistribution on consumer demand. In *Relevance and Precision*, Essays in Honour of Pieter de Wolff. Alphen aan den Rijn and Amsterdam: Samson/North-Holland Publishing Company, 123–138

1978

(1) On prediction. In *Proceedings Bicentennial Congress Wiskundig Genootschap*. MC Tract 100, Vol I: 123–132
(2) A function for size distribution of incomes: comment. *Econometrica*, 46: 459–460

1979

(1) De omvang van het betalingsverkeer en de omloopsnelheid van het geld: enige ramingen voor 1974–1977. In *Samenleving en Onderzoek*, Leiden: Stenfert Kroese: 209–224

1981

(1) The work money does – The transaction velocity of circulation of money in The Netherlands, 1950–1978. *European Economic Review*, 15: 307–326
(2) The volume of transactions and of payments in the United Kingdom, 1968–1977, *Oxford Economic Papers*, N.S., 33: 234–255

1982

(1) Het voorspelprobleem. *Maandblad voor Accountancy en Bedrijfshuishoudkunde*, 56: 356–364

1983

(1) Contanten: wit, zwart, onzichtbaar. *Kwantitatieve Methoden*, no. 9: 2–28
(2) Currency by denomination. *Economics Letters*, 12: 299–303
(3) De kosten van een huisvader (with Gusta Renes). *Kwantitatieve Methoden*, no. 10, 12–30
(4) Income distribution functions with disturbances (with M.R. Ransom). *European Economic Review*, 22: 363–372

1986

(1) The volume of transactions and the circulation of money in the United States, 1950–1979. *Journal of Business and Economic Statistics*, 4: 225–232
(2) Functional form of Engel curves for foodstuffs (with M.A.C. de Witte). *European Economic Review*, 30: 909–913
(3) *Econometric applications of maximum likelihood methods.* Cambridge: Cambridge University Press
(4) Herschatting van het Cramer autobezitsmodel (with A. van der Hoorn and A. Vos. *Verkeerskunde*, 37: 308–311
(5) De optimale coupure-samenstelling van de chartale geldhoeveelheid. *Bank- en Effectenbedrijf*, 35: 391–394
(6) Estimation of probability models from income class data. *Statistica Neerlandica*, 40: 237–250

Name index

Aaker, D.A. 205, 221
Abraham, B. 211, 221
Abrahamse, A.P.J. 185, 189, 191, 204
Ahluwalia, M.S. 165, 177
Aigner, D.J. 153, 162
Aitchison, J. 156, 163
Amemiya, T. 13
Anderson, T.W. 251, 253, 261
Ando, A. 47, 56
Ashley, R. 205, 221
Atkinson, A.B. 3, 8, 13, 163
Avery, R. 73, 98
Aviezer, S. 119

Bakker, B. 152, 163
Bartels, C.P.A. 154, 163
Barten, A.P. 23, 29, 181, 191, 204
Basmann, R.L. 251, 254, 261
Bell, W.R. 139, 147
Bera, A.K. 9, 13
Berry, A. 165, 166, 177
Bhattacharyya, M.N. 205, 211, 218, 219, 220, 221
Blackorby, C. 48, 56
Blundell, R.W. 3, 6, 7, 8, 9, 13
Boeschoten, W.C. 71, 78, 87, 93, 95, 98
Box, G.E.P. 135, 205, 209, 212, 222
Bretschneider, S.I. 205, 221
Brewer, K.W.R. 211, 221
Brotherton, T.W. 211, 221
Brown, J.A.C. 156, 163
Browne, F.X. 99, 101, 119
Brumberg, R. 47, 57
Brunner, K. 222
Buse, A. 200, 204

Caines, P.E. 211, 221
Carbone, R. 205, 221
Carman, J.M. 205, 221
Carvalho, J.L. 222
Champernowne, D.G. 154, 163
Chan, W.Y. 210, 221
Chen, C. 99, 101, 119
Chenery, H.B. 177
Chesher, A.D. 9, 13, 37, 43, 153, 163
Chow, G.C. 146, 147, 261
Clarke, D.G. 205, 221
Cleveland, W.P. 139, 148
Corsten, L.C.A. 222
Cragg, J.G. 3, 14
Cramér, H. 226, 245
Cramer, J.S. 3, 5, 14, 31, 45, 52, 55, 56, 69, 71, 72, 87, 89, 91, 98, 101, 119, 121, 130, 131, 147, 151–155, 157, 158, 160, 163, 164, 181, 185, 189, 191, 204, 220, 221, 222, 224, 225, 231–233, 236, 237, 245, 247, 263
Creedy, J. 154, 163
Cronin, D.C. 163

Dagum, C. 156, 163
De Crombrugghe, D. 189
Deaton, A.S. 4, 5, 8, 14, 15–21, 23, 24, 27, 29, 45, 48, 56, 57, 158, 163
De Vos, A.F. 131, 132, 135, 136, 148
De Witte, M.A.C. 266
De Wolff, P. 155, 163, 164, 265
Dhrymes, P.J. 261
Diewert, W.E. 16, 29
Dijkman, H.C. 147
Dolton, P. 31, 34, 43
Dronkers, J. 152, 163

Emami, J. 11
Esteban, J.M. 156, 158, 163

Fase, M.M.G. 71, 78, 87, 93, 95, 99, 101, 120, 205, 221, 222
Feige, E.L. 67, 69, 72, 73, 77, 89, 91, 92, 98
Fisher, I. 68–71, 78, 95–98
Fisher, J.M. 139, 148
Friedman, M. 67, 98, 181, 189

Gastwirth, J.L. 153, 163
George, V. 168
Georgantelis, S. 15
Gersh, W. 136, 137, 147
Gibrat, R. 156
Gilshon, A. 119
Goldberger, A.S. 153, 162
Goldfeld, S. 114, 120
Goldin, E. 80, 98, 99
Goldthorpe, J. 39, 43
Gomulka, J. 3, 8, 13
Gorman, W.M. 48, 49, 57, 158, 160, 161, 163
Gourieroux, C. 9, 14
Gradshteyn, I.S. 176, 177
Granger, C.W.J. 205, 208, 222, 224, 240, 246
Grether, D.M. 222
Griliches, Z. 246
Grupe, M.R. 139, 148
Guilkey, D.K. 206, 222
Guttman, A. 119

Hagenaars, A. 159, 164
Hall, R.E. 45, 57
Hanssens, D.M. 205, 211, 218, 220, 222
Harrison, A. 152, 159, 163
Harrison, P.J. 136, 147
Hartog, J.A. 154, 163
Harvey, A.C. 131, 132, 136, 137, 143, 147
Heertje, A. 164
Helmer, R.M. 211, 217–220, 222
Heijmans, R.D.H. 191, 264
Hillmer, S.C. 139, 147
Hood, W.C. 253, 254, 257–259, 261
Hope, K. 39, 43
Hotelling, H. 181
Houthakker, H.S. 45, 48, 56, 57
Hsiao, Ch. 206, 208, 222

Imbens, G. 31
Intriligator, M.D. 246

Irish, M. 4, 5, 8, 9, 13, 14

Jacobson, R. 205, 221
Jansen, H.W.M. 147
Jarque, C.M. 9, 13
Jenkins, G.M. 135, 209, 222
Jevons, W.S. 67
Johansson, J.K. 211, 217, 218, 222
Jorgensen, E. 175
Jorgenson, D.W. 56, 57
Jovanovic, B. 31, 32, 35, 36, 42, 43

Kakwani, N.C. 153, 155, 163
Kapoor, S.G. 205, 222
Kang, H. 208, 222
Kay, J.A. 3, 4, 8, 14, 55, 57
Keen, M.J. 3, 4, 8, 14, 55, 57
Keller, W.J. 257, 261
Kendall, M.G. 153, 161, 163, 181, 189
Kennett, M. 98
Keynes, J.M. 69, 96, 98, 223
Kimball, R.C. 97, 101, 120
Kitagawa, G. 136, 137, 147
Kiviet, J.F. 223, 238, 246
Klein, L.J. 261
Kloek, T. 153, 162, 163, 183, 189
Koerts, J. 185, 189, 191, 204
Kooiman, H.J. 147
Koopmans, T.C. 253, 254, 257–261
Korzec, M. 155, 164
Kroonenberg, N. 265
Kuhn, T.S. 247, 262
Kuznets, S. 165, 177

Lau, L.J. 56, 57
Lancaster, A. 31, 37, 43
Lange, O. 151, 164
Laurent, R. 69, 71, 72, 98
Lawson, R. 168
Leamer, E.E. 146, 147, 153, 164
Lebergott, S. 152, 159, 164
Lee, L.F. 9, 13
Ledolter, J. 205, 222
Leipnik, R.B. 254, 261
Leser, C.E.V. 56, 57
Liviatan, N. 54, 57
Longini, R.L. 205, 221
Lukacs, E. 194, 204
Lydall, H. 154, 164

Maddala, G.S. 156, 158, 164, 171, 172, 175–177, 200, 204
Madhok, P. 205, 222
Makepeace, G.H. 34, 43
Malinvaud, E. 226, 231, 246
Malthus, R. 96
Mandel, J. 78, 98
Mandelbrot, B. 154, 164
Manski, C.F. 80, 98, 99
Maravall, A. 205, 222
McDonald, J.B. 153, 156, 158, 164, 171, 172, 176, 177
Meghir, C. 3, 6, 7, 9, 13
Meltzer, A.H. 222
Mendis, L. 56
Mennes, L.B.M. 264
Mercer, J. 78
Merkies, A.H.Q.M. 151, 158, 160, 163, 164
Modigliani, F. 47, 56
Monfort, A. 9, 14
Morgan, H.M. 101, 120
Morris, C.N. 3, 4, 8, 14, 55, 57
Mount, T.D. 164
Muellbauer, J. 14, 15–19, 21, 23, 24, 27, 29, 45, 48, 49, 54, 56, 57
Mustert, G.R. 164

Na'aman, I. 119
Nerlove, M. 209, 222
Neudecker, H. 191, 247, 262
Neumann, G.R. 43
Newbold, P. 224, 240, 246

Osborn, D.R. 208, 222
Oshima, M.T. 165, 177

Pagan, A.R. 238, 246
Palda, K.S. 210, 218, 222
Palm, F.C. 221
Pareto, V. 151, 155, 164
Payne, L.C. 101, 120
Peled, S. 119
Pen, J. 153, 154, 164
Pfeffermann, O. 139, 148
Phillips, G.D.A. 15, 238, 246
Pierce, D.A. 139, 148
Pollock, D.S.G. 247, 251, 254, 262
Postelnicu, T. 222
Prais, S.J. 45, 56, 57
Primont, D. 48, 56

Pudney, S.E. 14

Quandt, R.E. 153, 164

Randall, E.B. 78, 98
Ransom, M.R. 151–153, 159, 164, 165, 171, 172, 177, 265
Ray, R. 8, 13
Reekers, G.M. 130, 264
Renault, E. 9, 14
Renes, G. 265
Ricardo, D. 96
Ridder, G. 223
Rivlin, A.M. 151, 164
Rothenberg, T.J. 22, 29
Rothschild, M. 154, 164
Rubin, H. 251, 253, 260, 261
Ruud, P.A. 238, 246
Russell, R.R. 48, 56
Ryzhik, I.M. 176, 177

Sahota, G.S. 157, 164
Salem, A.B.Z. 164
Salemi, M.K. 206, 222
Scharf, W. 261, 262
Schmalensee, R. 205, 221
Selden, R. 67, 98
Sen, A.K. 154, 164
Sethi, S.P. 211, 221
Shephard, R.W. 16, 29
Sims, C.A. 208, 222
Singh, S.K. 156, 158, 164, 171, 172, 175–177
Sirkis, S. 119
Smith, J.T. 153, 163
Soltow, L. 165, 177
Spindt, P. 73, 98
Stern, N.H. 3, 8, 13
Stevens, C.F. 136, 147
Stiglitz, J.E. 154, 164
Stone, J.R.N. 15, 17, 29
Stuart, A. 153, 161, 163

Theil, H. 59, 62, 63, 181, 187, 219, 222, 247, 251, 254, 262
Tiao, G.C. 205, 209, 212, 222
Tinbergen, J. 153, 154, 164, 223
Todd, P.H.J. 132, 136, 137, 147
Tolk, P.C. 147
Trognon, A. 9, 14
Tsay, R.S. 209, 222

Umashankar, S. 205, 222

Van Daal, J. 158, 160, 164
Van Dijk, H.K. 153, 162, 163
Van Eck, W. 159, 164
Van de Gevel, F.J.J.S. 221
Van der Hoeven, D. 99, 119
Van der Hoorn, A. 266
Van Nieuwkerk, M. 99, 119, 121
Van Praag, B.M.S. 159, 164, 264
Van Renswoude, G.J. 161, 164
Van der Sar, N.L. 164
Van der Wijk, J. 156
Venekamp, P. 163
Vos, A. 266

Walker, I. 8, 13
Wallis, K.F. 210, 221
Wei, W.W.S. 211, 222
Weiskoff, R. 165, 166, 177
Westergaard-Nielsen, N.C. 43
White, H. 236, 237, 246
Williamson, P. 34, 43
Wold, H.O.A. 225, 246
Wolfe, L.A. 78
Working, M. 56, 57, 60, 62, 63
Wu, S.M. 205, 222

Young, A.H. 135, 148

Zellner, A. 208, 222, 223, 241, 246, 247, 262
Zhang, W. 15

Subject index

A
Adaptive behaviour 205
Adding-up condition 60
Addilog model 60
Aggregation 206
AIC (Akaike Information Criterion) 146
AIDS (Almost Ideal Demand System) 15–19, 21, 22, 24–28, 56
Averge life time of currency notes 71, 73

B
Best linear predictor 226
Best linear unbiased estimator 183, 186
Bivariate model 3, 4, 184
Bivariate (normal) distribution 6

C
Calendar variations: effects and models 131, 135, 137, 139, 140, 142
Cash loop 70, 71, 89
Cauchy distribution 240, 245
Causality 223
Censored univariate normal model 3
Census X-11 method 135
Central limit theorems 157
Chi-squared distribution 103
Coefficient of determination 181–183, 240, 245
Coefficient: Gini's – see Gini, Theil's inequality-see Theil.
Coefficient of variation 160
Commodity expenditure 3
Complementary subsystem 258
Concentrated likelihood function 258, 259
Conditional distribution 227, 229
Constant-elasticity demand model 60
Consumption theory 59
Convolution 157
Cowles commission 247
Curve, Engel – see Engel, Kuznets – see Kuznets, Lorenz – see Lorenz.

D
Dalton-Pigou condition 154
Data generating process (D.G.P.) 224, 234, 237
Demographic characteristics 5, 11
Denominational mix (of currency) 99
Denomination-specific demand (equation) 101, 103, 119
Density function 175
Discrete variables 233
Double-hurdle model 3

E
Econometric modelling 223
Economic growth 165, 166
Edgeworth expansion 22
Efficient payment 101, 103, 111
Eigenvalues 192, 193, 195, 198, 202
Elasticity of (intertemporal) substitution 45, 51
Empirical moments 250
Endogenous variable 248
Engel curve (estimation, model, aggregation) 3–5, 7–9, 16, 45, 51, 52, 55, 56, 155, 158
Engel's law 48
Equation of exchange 68, 71, 95, 96
Equivalence scale 45
Errors-in-variables model 253
Exogenous variable 248
Exponential distribution 103

F
Family expenditure survey (U.K.) 8, 11
Frequency-of-purchase model 5

G
Galton-Yale specification (approach) 226, 228, 230, 231, 236
Gamma distribution 103, 156, 158–160
Gauss-Fisher specification (approach) 226, 228, 231, 237, 238
Gini's coefficient (constant) 154
Goodness of fit (measure) 182, 185, 189, 191, 200
Gorman's Theorem 49, 55
Granger-Sims causality 205, 208, 214

H
Hazard (function) 33, 38
Heteroskedasticity 200
Histogram 172
Homogeneity restrictions 16, 19, 21–24, 27
Homoskedasticity 241
Homotheticity (within period) 49
Homothetic utility function 14
Hypergeometric function 176

I
Identifiability 249
IMF 122, 126
Imports (world) 121
Income distribution 165
Income-expenditure (model, equation) 4, 8
Income redistribution 155, 158, 160
Independence test 9
Indirect utility function 49
Individual characteristics 3, 6
Individual survey data 3, 6
Inferior goods 22
Iterative seemingly unrelated (estimation by) 19

J
Jensen's inequality 240
Job search 32
Job separation 31

K
Kalman filter 131, 136, 142, 146
Kaplan-Meier estimate 37
Khintchine's Law 196

Koyck model 205
Kuznets curve 165

L
Lagrangean multiplier 251
Lagrange multiplier test 9, 22, 24, 26
Latent root 253
Latent variable 5, 6
Law, Engel's – see Engel, Khintchine's – see Khintchine, Pareto's – see Pareto, Shephard's – see Shephard.
Least-squares regression (estimation by) 251, 153
Life-cycle (consumption) hypothesis 46
Lifetime of currency notes 71, 73
Likelihood function 251
Likelihood-ratio test 22, 26
Limited-information maximum-likelihood estimation 253
Linear least-squares estimation 226, 233
Lognormal distribution 59, 60, 156, 159
Longitudinal analysis 166
Lorenz curve 154, 169, 171, 172, 176
Lydia Pinkham Medicine Co. 206

M
Maximum-likelihood estimation 21, 153, 251
Maximum-likelihood estimator 251
Minimum chi-square (estimation) 153
Minimum variance-ratio method 254
Misspecification test (analysis) 224, 237
Model, bivariate – see bivariate, double-hurdle see double-hurdle, Koyck – see Koyck, overidentified – see overidentified, Rotterdam – see Rotterdam, simultaneous – see simultaneous, Tobit – see Tobit.
Moments (estimation by) 153, 250
Multiequation subsystem 254
Multinomial loglikelihood function 172
Multivariate model 184
Multivariate normal distribution 59, 226
Multivariate stochastic time series 205

N
Nearly (Almost) Ideal Demand System – see AIDS
Non-inferior goods 155
Non-linear least-squares regression 230
Normal distribution 59, 196, 226

O
Opportunity cost 104, 107, 112, 113
(Optimal) job search 32
Overidentification 250, 251

P
Pareto distribution 151–156
Pareto's Law 154, 157
PIGLOG class 16
Poincaré's separation theorem 193
Population correlation coefficient 185, 192
Population moments 250
Probability limit (plim) 185–187, 194, 196, 197, 199–203, 227, 229, 239, 242
Procrustes transformation 139, 140, 142, 146

Q
Qualitative variables 233
Quantiles (estimation by) 153, 172

R
R^2 181, 183–188, 191, 192, 196, 197, 199–201, 203
Reduced form (parameters) 249, 151
Risk neutrality 33
Rotterdam model 15–18, 21, 24, 26–28

S
Sample correlation coefficient 192
Separability (intertemporal weak, intertemporal additive) 49
Shephard's lemma 50
Simultaneous model 247
Singh-Maddala distribution 158, 162, 171, 172, 175, 176
Single-equation estimator 250
Social welfare function 154
Spurious regression 224, 238, 240, 241
Squared correlation coefficient 181

Stone's price index 15, 27
Structural parameters 249, 252
Symmetry restrictions 16, 19, 21–23, 26, 27
System of demand equations 59

T
Theil's inequality coefficient 154
Theorem, Gorman's – see Gorman,
   Poincaré's – see Poincaré.
Three-stage least squares (estimation by) 247
Time series (non-experimental economic) 223, 230, 239
Tobit model 3, 4
Trade (world) 121, 122
Transactions 101, 103, 131
Translog system 56
Trends and seasonality 131–134
Two-stage least squares (estimation by) 247

U
U.K. Family Expenditure Survey 8, 12
United Kingdom Accounts Blue Book 19
Utility function 59

V
Validation checks 224
Velocity of currency circulation 67–69, 71, 73, 76, 79, 81, 86, 97

W
Wald test 22, 27
World balance of payments 121, 125, 126
World exports 121
World imports 121
World trade 121, 122

Z
Zero expenditure 3, 4, 8